Computational Methods and
Experimental Measurements XX

WITPRESS

WIT Press publishes leading books in Science and Technology.
Visit our website for new and current list of titles.
www.witpress.com

WITeLibrary

Home of the Transactions of the Wessex Institute.
Papers published in this volume are archived in the WIT eLibrary in volume 130 of
WIT Transactions on Engineering Sciences (ISSN 1743-3533).
The WIT electronic-library provides the international scientific community with
immediate and permanent access to individual papers presented at WIT conferences.
Visit the WIT eLibrary at www.witpress.com.

TWENTIETH INTERNATIONAL CONFERENCE ON
COMPUTATIONAL METHODS AND EXPERIMENTAL MEASUREMENTS

CMEM XX

CONFERENCE CHAIRMEN

Santiago Hernández
University of A Coruña, Spain
Member of WIT Board of Directors

Giovanni Maria Carlomagno
University of Naples "Federico II", Italy

LOCAL ORGANISER

Guido Marseglia
University of Seville, Spain

INTERNATIONAL SCIENTIFIC ADVISORY COMMITTEE

ORGANISED BY

Wessex Institute, UK
University of Naples "Federico II", Italy

SPONSORED BY

WIT Transactions on Engineering Sciences

International Journal of Computational Methods and Experimental Measurements

WIT Transactions

Wessex Institute
Ashurst Lodge, Ashurst
Southampton SO40 7AA, UK

K. Dorow Pacific Northwest National Laboratory, USA

W. Dover University College London, UK

C. Dowlen South Bank University, UK

J. P. du Plessis University of Stellenbosch, South Africa

R. Duffell University of Hertfordshire, UK

A. Ebel University of Cologne, Germany

V. Echarri University of Alicante, Spain

K. M. Elawadly Alexandria University, Egypt

D. Elms University of Canterbury, New Zealand

M. E. M El-Sayed Kettering University, USA

D. M. Elsom Oxford Brookes University, UK

F. Erdogan Lehigh University, USA

J. W. Everett Rowan University, USA

M. Faghri University of Rhode Island, USA

R. A. Falconer Cardiff University, UK

M. N. Fardis University of Patras, Greece

A. Fayvisovich Admiral Ushakov Maritime State University, Russia

H. J. S. Fernando Arizona State University, USA

W. F. Florez-Escobar Universidad Pontifica Bolivariana, South America

E. M. M. Fonseca Instituto Politécnico do Porto, Instituto Superior de Engenharia do Porto, Portugal

D. M. Fraser University of Cape Town, South Africa

G. Gambolati Universita di Padova, Italy

C. J. Gantes National Technical University of Athens, Greece

L. Gaul Universitat Stuttgart, Germany

N. Georgantzis Universitat Jaume I, Spain

L. M. C. Godinho University of Coimbra, Portugal

F. Gomez Universidad Politecnica de Valencia, Spain

A. Gonzales Aviles University of Alicante, Spain

D. Goulias University of Maryland, USA

K. G. Goulias Pennsylvania State University, USA

W. E. Grant Texas A & M University, USA

S. Grilli University of Rhode Island, USA

R. H. J. Grimshaw Loughborough University, UK

D. Gross Technische Hochschule Darmstadt, Germany

R. Grundmann Technische Universitat Dresden, Germany

O. T. Gudmestad University of Stavanger, Norway

R. C. Gupta National University of Singapore, Singapore

J. M. Hale University of Newcastle, UK

K. Hameyer Katholieke Universiteit Leuven, Belgium

C. Hanke Danish Technical University, Denmark

Y. Hayashi Nagoya University, Japan

L. Haydock Newage International Limited, UK

A. H. Hendrickx Free University of Brussels, Belgium

C. Herman John Hopkins University, USA

I. Hideaki Nagoya University, Japan

W. F. Huebner Southwest Research Institute, USA

M. Y. Hussaini Florida State University, USA

W. Hutchinson Edith Cowan University, Australia

T. H. Hyde University of Nottingham, UK

M. Iguchi Science University of Tokyo, Japan

L. Int Panis VITO Expertisecentrum IMS, Belgium

N. Ishikawa National Defence Academy, Japan

H. Itoh University of Nagoya, Japan

W. Jager Technical University of Dresden, Germany

Y. Jaluria Rutgers University, USA

D. R. H. Jones University of Cambridge, UK

N. Jones University of Liverpool, UK

D. Kaliampakos National Technical University of Athens, Greece

D. L. Karabalis University of Patras, Greece

A. Karageorghis University of Cyprus

T. Katayama Doshisha University, Japan

K. L. Katsifarakis Aristotle University of Thessaloniki, Greece

E. Kausel Massachusetts Institute of Technology, USA

H. Kawashima The University of Tokyo, Japan

B. A. Kazimee Washington State University, USA

F. Khoshnaw Koya University, Iraq

S. Kim University of Wisconsin-Madison, USA

D. Kirkland Nicholas Grimshaw & Partners Ltd, UK

E. Kita Nagoya University, Japan

A. S. Kobayashi University of Washington, USA

D. Koga Saga University, Japan

S. Kotake University of Tokyo, Japan

Computational Methods and Experimental Measurements XX

EDITORS

Santiago Hernández
University of A Coruña, Spain
Member of WIT Board of Directors

Giovanni Maria Carlomagno
University of Naples "Federico II", Italy

Guido Marseglia
University of Seville, Spain

WITPRESS Southampton, Boston

Editors:

Santiago Hernández
University of A Coruña, Spain
Member of WIT Board of Directors

Giovanni Maria Carlomagno
University of Naples "Federico II", Italy

Guido Marseglia
University of Seville, Spain

Published by

WIT Press
Ashurst Lodge, Ashurst, Southampton, SO40 7AA, UK
Tel: 44 (0) 238 029 3223; Fax: 44 (0) 238 029 2853
E-Mail: witpress@witpress.com
http://www.witpress.com

For USA, Canada and Mexico

Computational Mechanics International Inc
25 Bridge Street, Billerica, MA 01821, USA
Tel: 978 667 5841; Fax: 978 667 7582
E-Mail: info@compmech.com
http://www.witpress.com

British Library Cataloguing-in-Publication Data

A Catalogue record for this book is available
from the British Library

ISBN: 978-1-78466-425-1
eISBN: 978-1-78466-426-8

ISSN: 1746-4471 (print)
ISSN: 1743-3533 (on-line)

The texts of the papers in this volume were set individually by the authors or under their supervision. Only minor corrections to the text may have been carried out by the publisher.

No responsibility is assumed by the Publisher, the Editors and Authors for any injury and/or damage to persons or property as a matter of products liability, negligence or otherwise, or from any use or operation of any methods, products, instructions or ideas contained in the material herein. The Publisher does not necessarily endorse the ideas held, or views expressed by the Editors or Authors of the material contained in its publications.

Preface

This volume includes most of the contributions offered by several specialists at the 20th International Conference on Computational Methods and Experimental Measurements (CMEM21), held in 2021 and delivered online.

The sequence of these conferences appears exclusively in the related field and it presently derives from a forty-year tradition. Indeed, this series was conceived in 1981 by our friend, the late Professor Carlos Antonio Brebbia, who we will never forget, and has been reconvened by meetings on board the Queen Elizabeth II Ocean Liner (1984); Porto Carras, Greece (1986); Capri (1988); Montreal (1991); Siena (1993); Capri (1995); Rhodes (1997); Sorrento (1999); Alicante (2001); Sani Beach, Greece (2003); Malta (2005); Prague (2007); the Algarve (2009); the New Forest, UK (2011); A Coruña, Spain (2013); Opatija, Croatia (2015); Alicante, Spain (2017) and Seville, Spain (2019). This is the first time that the meeting has had to be delivered online.

Key objective of the conference has always been to offer to the scientific and technical community an international atmosphere to analyze the interaction between computational methods and experimental measurements with their corresponding features, the foremost concern and significance being devoted to their mutual and beneficial merging.

The constant development of numerical procedures and computer efficiency, coupled with their decreasing costs, have generated an ever increasing growth of computational methods that are currently exploited both in a persistently expanding variety of science and technology subject matters, as well as in our daily lives.

Nevertheless, it must be observed that, even if these procedures are continuously growing and getting more reliable, it is still required to have an accurate validation of some of them, particularly the most complex ones. This can be only achieved by performing dedicated and precise experimental tests. The currently available experimental techniques allow such tests to be performed but, since they are obviously evolving to be more and more intricate and elaborate, often both rigs running as well as data acquisition can be only performed with the help of computers.

Finally, it has also to be pointed out that, for many of the modern experimental techniques, the amount of obtained data can be enormous so that processing by means of numerical methods becomes unavoidable.

The Editors are very grateful to all the authors for their valuable contributions and to the Members of the International Scientific Advisory Committee, as well as some other colleagues, for their help in reviewing the papers which appear in this volume.

The Editors, 2021

Contents

SECTION 1
COMPUTATIONAL AND
EXPERIMENTAL METHODS

DESIGN OF EXPERIMENTS PLATFORM FOR ONLINE SIMULATION MODEL VALIDATION AND PARAMETER UPDATING WITHIN DIGITAL TWINNING

MADHU SUDAN SAPKOTA[1], EDWARD APEH[1], MARK HADFIELD[1],
ROBERT ADEY[2] & JOHN BAYNHAM[2]
[1]Faculty of Science and Technology, Bournemouth University Poole, UK
[2]CM BEASY Ltd., UK

ABSTRACT

The process of developing a virtual replica of a physical asset usually involves using standardized parameter values to provide simulation of the physical asset. The parameters of the virtual replica are also continuously validated and updated over time in response to the physical asset's degradation and changing environmental conditions. The parametric calibration of the simulation models is usually made with trial-and-error using data obtained from manual survey readings of designated parts of the physical asset. Digital Twining (DT) has provided a means by which validating data from the physical asset can be obtained in near real time. However, the time-consuming process of calibrating the parameters so the simulation output of the virtual replica matches the data from physical asset persists. This is even more so when the calibration of the simulator is performed manually by analysing the data received from the physical system using expert knowledge. The manual process of applying domain knowledge to update the parameters is error prone due to incompleteness of the knowledge and inconsistency of the validation/calibration data. To address these shortcomings, an experimental platform implemented by integrating a simulator and a scientific software is proposed. The scientific software provides for the reading and visualisation of the simulation data, automation of the simulation running process and provide interface of the relevant validation and adaptive algorithmics. This comprehensive integrated platform provides an automated online model validation and adaptation environment. The proposed platform is demonstrated using BEASY – a simulator designed to predict protection provided by a cathodic protection (CP) system to an asset, with MATLAB as the scientific software. The developed setup facilitates the task of model validation and adaptation of the CP model by automating the process within a DT ecosystem.
Keywords: model adaptation, Digital Twin, cathodic protection, BEASY, software integration.

1 INTRODUCTION

In undertaking the structural health monitoring task, engineers use simulation tools to predict the risk to the structure from degradation mechanisms such as corrosion and cracking. More often, the parametric simulation models derived from the simulators on calibration are used to predict the location of the damage and its severity. Parameters as the model independent variables are model's input describing the properties of the materials and the environmental conditions the structure experiences. These model input variables cannot be directly measured due to structural complexity in most of the situations. Parameter setting for the model during realisation of a virtual replica of an existing physical structure thus relies upon the structural response-data obtained from sensors or survey. During such parametric calibration, the best set of parameters are found by correlating the model output to the available measurements from the physical system [1].

The traditional approach of calibration when performance validating data are available from the structure is based on a trial-and-error procedure involving manual iterative analyses and modifications. This task is normally performed by the engineers using their experience to choose the appropriate values until good correspondence is obtained (Fig. 1). As the parameters change with time, the process is repeated over the life of the structure based upon

WIT Transactions on Engineering Sciences, Vol 130, © 2021 WIT Press
www.witpress.com, ISSN 1743-3533 (on-line)
doi:10.2495/CMEM210011

the requirement determined by expert and availability of the dataset. This results in significant computational efforts when performing model calibration/updating for complex models and in a repetitive manner.

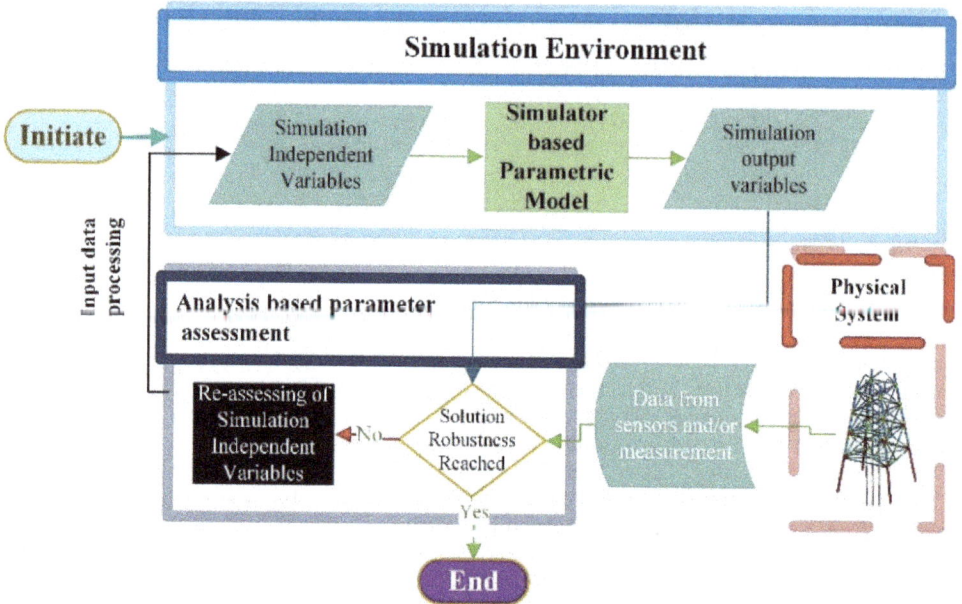

Figure 1:   The parametric simulation model parameter updating for performance enhancement with output analysis (trial-and-error method).

The design optimisation [2] algorithms could be used to search the parameter space for the optimal set of parameters. However, the challenges like selecting the relevant optimisation algorithm, and implementing to the model still exist. These challenges are still relevant in the situations where commercial simulators are used for model building, but the platform for automated experimentation and analysis is absent. A platform that combines an optimisation software with a simulator can form a basis for decision-making for virtual replica [3]–[5]. Such an integration is more relevant and essential when two major tasks of modelling i.e. geometrical modelling (CAD design) and numerical simulation are provided by different software [6].

The virtual replica of an existing physical counterpart is recognised as Digital Twin (DT) in recent years. DT has been proposed with self-adaptive simulation characteristics and predictive capability [7]. Analytical supports to provide self-adaptation are anticipated incorporated within DT [8]. However, while enabling DT provided with the analytical features encapsulated within it, the features are not necessarily available under same software platform as for simulation. Therefore, the software integrated platform is still an important aspect for a DT model to enable a self-adaptive DT. The requisite of such experimental platform is to exploit the relevant features from available tools and/or software on enabling simulator-based Digital Twin.

This paper proposes software integration to have an experimental design platform required for continuous experimentation and analysis of the simulator-based prediction, while enabling and maintaining a DT. The supporting software representing the server of platform is either enriched with the analytical capability or analytical tool(s) interfacing capability, while simulating software is major tool for process simulation. This software integration platform provides a step towards establishing the core aspect of DT i.e. self-adaptation, leveraging the analytical aspects within DT.

2 SIMULATOR BASED DIGITAL TWIN ENABLING

2.1 Simulator based modelling

The use of previously validated model to the conceptual level assists in the design of not just one simulation model but many within the problem domain [9], [10]. The parametric simulation model building in multiple domains is already facilitated by the availability of commercial process simulators (simulating-software). Such simulators are generally based on current scientific understanding (physics of phenomena), often involving numerical approximation of differential equations, and they are implemented in complex computer programs [11]. The numerical approximation methods (e.g. Finite Element Method, Boundary Element Method) adopted for process simulation usually involves three major steps [12], [13]: (i) geometrical modelling of the structure(s), (ii) meshing of the geometry, and (iii) computational approximation of the Partial Differential Equation (PDE) solution. In tools supported modelling, steps (i) and (ii) are performed with CAD and meshing software tools, respectively, and then step (iii) requires a numerical solver. Commercially available simulators (e.g. AKSELOS, ANSYS, BEASY) provide the functionality of CAD modelling, meshing, as well as the behavioural process simulation i.e. numerical approximation of PDE.

When the commercially available simulator(s) provides the primary modelling requirements, the major efforts in realising virtual replica of physical asset are focused on determining model independent variables (parameters) such as material and environmental related parameters. The trial-and-error (Fig. 1) based traditional method during parametric calibration is inefficient and introduces time-delays due to manual performance, also has the drawbacks of not having the ability to calibrate multiple parameters simultaneously, and do not guarantee the best results. The simulation environment though enriched with core process simulation solver, thus requires analytical support that overcomes the drawbacks of traditional methods while tailoring the parametric model.

Experimental design, sensitivity analysis, and design (parameter) optimisation are widely adopted techniques within the systematic procedures of model performance enhancement [14]. Design optimisation is an approach that combines mathematical optimization algorithms with parametric simulation model to search the design (parameter) space for the optimal solution [2]. Gradient-based and non-gradient based algorithms are used during parameter optimisation [15], [16] for reaching the best fit. However, the challenges on selecting relevant techniques/algorithms still exists and remains time-consuming when selection and/with implementation to model is performed manually.

Automation of the analytical task by integrating online analytical environment to the simulation environment is the next step on addressing the pre-mentioned drawbacks. Correspondingly, the design optimisation functionality anticipated while determining best set of parameters can be facilitated using pre-available supporting tools (commercial-scientific software).

2.2 Digital Twin concept on modelling

The DT concept is a new paradigm of interest in the field of modelling and simulation having online simulation as a core functionality of the system employing seamless assistance [17]. The definitions on DT are evolving rapidly in recent years as more insights are added to its features/potentials. Henceforward, DT is being preferred over the model in the context of adaptive simulation [18]. Moreover, the aspects of self-parameterisation and self-adaptation allowing DT to resemble its counterpart physical twin are of high importance to realize the potential of artificial intelligence (AI) within DT [8], [19]. Also, a digital twin concept is not only about a single behavioural simulation, mostly involved multiple process co-simulations representing different behavioural phenomena. This highlights the DT as collaborative models to provide a comprehensive representation of the system's simultaneously occurring multiple dynamic phenomena. Going along with the concepts on DT, while enabling DT as high-fidelity simulation model provided with self-analysis features or DT with co-simulations, software integration [7] is anticipated.

Despite the limitations of not having standardized concept on DT, one can reap the benefits that the DT could provide, like Digital Twinning on having validating data from the physical asset in near real time. Thus, software integration requisite for model performance enhancement is preferably undertaken within Digital Twin concept, not only to utilise the benefits DT provides, but also to standardise the DT's potential(s) and requirements.

3 DESIGN OF EXPERIMENTS PLATFORM WITH SOFTWARE INTEGRATION

The implementation of open-source or commercial software (tools) provides the analytical assistance during model calibration/adaptation i.e. implementing adaptive algorithms to the parametric models [3]–[5]. Such supporting tools when used in collaborative workflows with the model makes the simulation easier to compile, test, analyse, and suggest changes to model's input. On the other hand, adopting DT concept means incorporating features like potential to handle data, perform experimentation, and implement adaptive algorithm together with simulator to have a robust predictive tool. For this, experimental platform providing the integration between the major tool (simulator(s)) and the supporting tool(s), with automated data and control signal flow between them is proposed within Digital Twinning concept.

The scientific software should either include tool(s) (Fig. 2) or at least provide the interface to the external tool(s) for analytical supports including design optimisation. This support then would be utilised within the platform where scientific software (server) serves the simulator as its client [20]. Before this, the server has the role of input–output data management like filtering and data mapping, preparing dataset for the simulator and the analytical tool(s). Towards, analytical aid on model performance enhancement, it provides a design of experiments platform for experimentation and analysis using designated internal/externally interfaced algorithms. The server software similarly provides the opportunity for the visualisation of the data to denote the state and discrepancies in the performance of the model from the actual system. Likewise, automated data flow between the platform and the data-server allows retro-perspective analysis when corresponding time-series data are available into the data server. In addition, the platform should provide model performance validating criterion/algorithm separately or integrated to the platform, as adaptation task cannot be carried out without validation task synchronised [21].

While enabling DT and/or maintaining it over the lifetime, the scientific software-based integrated platform forms the basis of automation of the parameter updating required for model performance validation and enhancement.

4 CASE STUDY: BEASY BASED CATHODIC PROTECTION MODELLING AND ITS ENHANCEMENT

4.1 Background

Cathodic protection (CP) is most frequently used for the protection of underground or underwater (sea water) metallic infrastructures from corrosion. The design of the CP system depends upon assumptions (often referred to as design rules) about the performance of other protection measures such as coatings, in particular the rate at which coatings are to be assumed to degrade over the life of the structure. The performance of the CP system can be evaluated and optimised based on the design rules using a CP simulation model which predicts year by year the protection potentials, the depletion of the anodes, and in the case of Impressed-Current-Cathodic-Protection (ICCP), the current to be required by the system over its life [22].

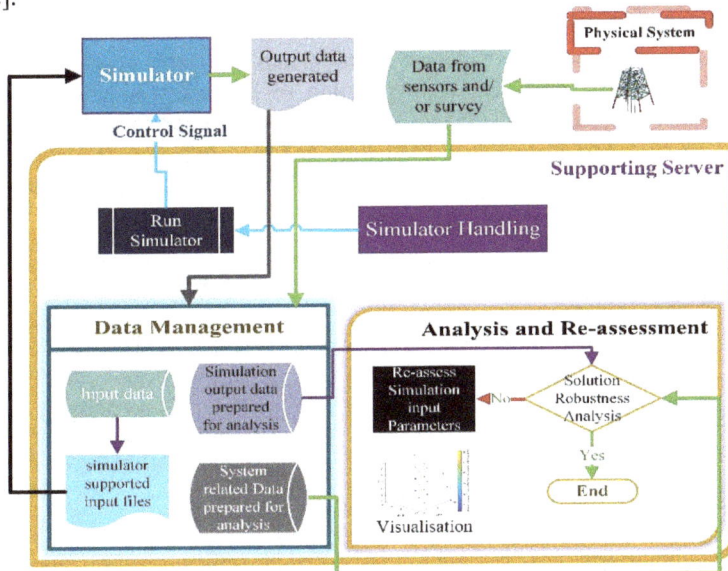

Figure 2:   Integration of server and simulator for automation of the process of data management and analytical support required on simulator-based DT enabling.

While this type of simulation provides valuable information to the design engineer by confirming that the required protection will be achieved. In reality, the actual performance of the CP system will be often different as coatings for example often degrade at different rates to that described in the design rules, environmental conditions may vary, the "as-built" structure may be different and changes and retrofits are made over time. Integrating the CP data collected during the routine inspection surveys with a CP simulation model on calibrating/adapting the model to match the inspection data enables a "digital twin" of the structure [23]. In this way, the digital twin represents the behaviour of the structure and the CP system at the time the inspection survey was performed. This then provides the ability to predict the present and future protection for all parts of the structure.

By repeating the process with each new inspection report the engineer can monitor the differences between the model predictions and survey data systematically to assess current "health" of the structure, identify anomalies, predict and plan for future risks, optimise the inspection strategy and provide early identification of problems which will require actions.

4.2 BEASY as a simulator for cathodic protection model

BEASY tool – a commercial parametric simulator designed to simulate the behaviour of galvanic corrosion problems and cathodic protection designs is adopted to represent a virtual replica of a CP system.

The CP simulation model is based on the prediction of the distribution of electrical potential and protection current density on the electrode surfaces as well as at the corresponding points. The calculation of electric potential and current density distribution is based on the solution of the well-known Laplace partial differential eqn (1).

With assumption that electrolyte is homogeneous

$$-\nabla(k\,\nabla\varphi) = 0, \tag{1}$$

where k = electric conductivity, φ = electric potential and ∇ is Nabla operator.

BEASY-CP provides numerical approximation of Laplace's equation obtained with Boundary Element Method (BEM) [11], [22] for steady state corrosion. Also, commercial BEASY tool facilitates on geometrical modelling and meshing. The data about geometry, meshing, and materials and surrounding related parameters data can be exported to text files and feed to the solver for numerical approximation (Fig. 3).

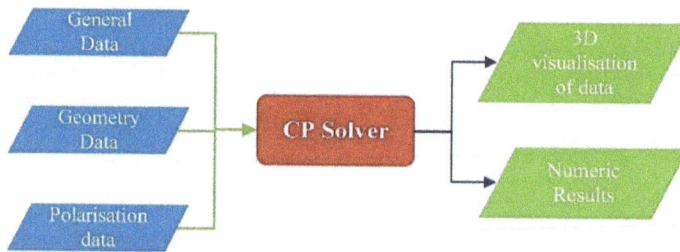

Figure 3: Inputs–outputs for BEASY tool based cathodic protection simulation.

The numeric results provided by the simulation are analysed with data collected during the routine inspection surveys of physical CP system. Calibration is made with repetitive analysis and updating of the input parameter related value into the input supported files before to run next simulation and get new data for analysis.

A CP simulation model of a marine structure (Fig. 4) protected by sacrificial anodes was built using the BEASY tool. The model parameters required by the cathodic protection model for the CP system of structure (Fig. 4) are:

- Polarisation behaviour: The relationship between potential and current density represents the electrode kinetics of the metal in the seawater. It provides the boundary condition while solving the numerical problem.
    - Polarisation curve for Material 1 of the structure (Fig. 4).
    - Polarisation curve for Material 2 of the structure (Fig. 4).
- Conductivity/resistivity: Surrounding medium/material related.
    - Sea water related conductivity (Siemen/m).
    - Sea-bed related conductivity (Siemen/m).

WIT Transactions on Engineering Sciences, Vol 130, © 2021 WIT Press
www.witpress.com, ISSN 1743-3533 (on-line)

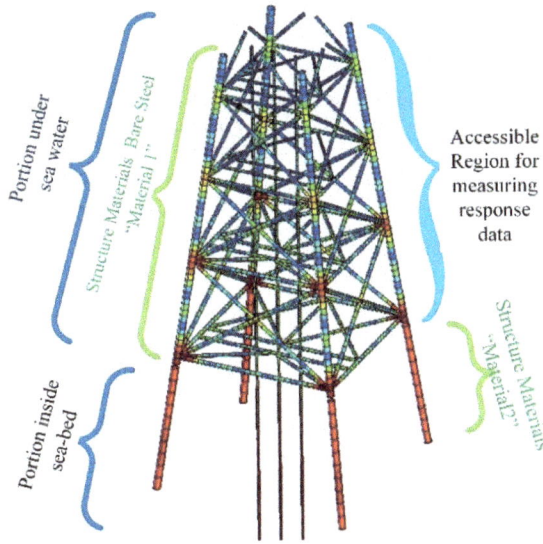

Figure 4:   The project adopted geometrical model of marine structure protected with sacrificial anodes [21].

As the polarisation curves for representing the polarisation behaviour are graphical representation (Fig. 5) and dynamics with time, a quantitative representation should be established for readjustment of the polarisation data. To address the possible change in polarisation behaviour, the curve transformation value (expanding or squeezing factor) is taken as a variable (parameter) keeping the curve constant obtained from design rule. This parameterisation concept can be understood as a modification of the diffusion limiting current in the polarisation behaviour of the materials involved. The transformation vector or parameter is termed as "p-value" in this case study.

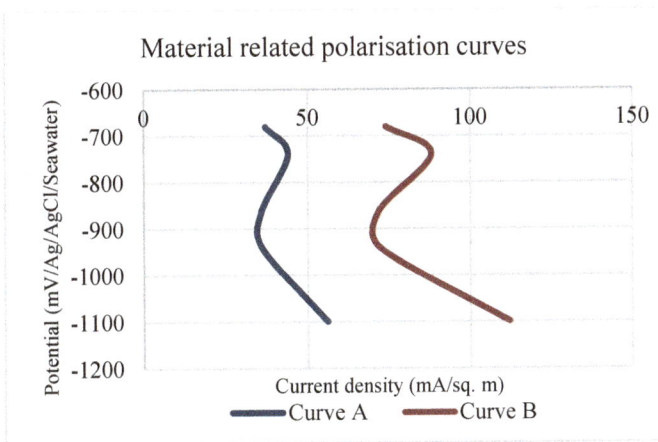

Figure 5:   Two different polarisation curves that can be transformed from one to other with some transformation value [21].

4.3 Software integration with MATLAB [24] as a server

Even though the simulation task is mostly dependent upon the simulator, the scientific software in the design of experiments platform remains as the server. The extensive data analysis, plotting capability and the availability of different optimisation algorithm, with the MATLAB enables completion of assess–modify–check loops in reduced computational time. Optimisation tool within MATLAB such as "fmincon" and "fminunc" [24] likewise can be used for constrained and unconstrained optimisation problem respectively. MATLAB – a scientific software granted with these qualities is selected as the Server for the integration platform to enable DT for a CP system.

Furthermore, data reading and modifying from/to the input–output files to the simulation solver can be performed with the server software or using other open-source tools (e.g. python based). MATLAB as server also provides the integrating ability to such tools. In this case study, python-based codes were incorporated to facilitate the data transfer and update in files shared between server and the simulator.

4.4 Tools integration for optimisation based parameter updating and analysis

4.4.1 Tool for optimisation-based parameter calibration/adaptation
For the task of adaptation of the CP model with parameter fitting, a gradient based algorithm "quasi-newton" is chosen. The reason for choosing gradient based approach as opposed to other optimisation techniques, such as genetic algorithms or neural networks, is that the problem space of CP model is mostly monotonic. Optimised values of parameters can, therefore, be found quickly with gradient-based algorithms, i.e. in a lower iteration(laps) count and consequently in a lower amount of time. On this basis, unconstrained optimisation tool "fminunc" within MATLAB is utilised and supported with "quasi-newton" [20] algorithm.

4.4.2 CP model's performance validating data
The data types that can be practically obtained from the structure and from the CP model simulation run as well are: (a) surface potential (mV), (b) normal current density (mA/m^2) and (c) electric field (mV/m) [22]. However, the data dependency for validation depends upon the complexity of the model.

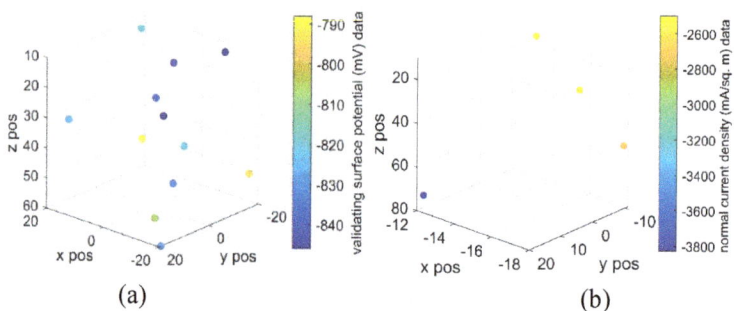

Figure 6: The validation data obtained after simulation run from the reference model. (a) Surface potential; and (b) Normal current density, from selected validating position of structure's surface (Fig. 4).

At this stage, validating data are generated from a virtual reference model provided with different parameter values considering the possible polarisation behaviour in future than to the parameter values set for the initial primary model whose value(s) mostly depends upon the design rules. Two types of validating data are considered surface potential (mV) and normal current density (mA/m^2) and the validating data positions count are 12 and 4 respectively from the structure's surface (Fig. 6).

4.4.3 Objective function for optimisation problem

Reducing the discrepancies between the validating and response data from simulation is the goal of optimisation. Assuming two different validating datatypes (Section 4.4.2), Normalised Mean Square difference between validating and model output data with weightage constant (2:1) provided for both data types, is used as an Objective function.

4.4.4 Initial stage primary model and parameters value

The highly sensitive parameters are the focus of in this case study, which is why the other parameters' values are kept constant. It is easily obtained from sensitivity analysis that the parameters more sensitive to response data are the "p-value of Material 1 related polarisation curve" and "Sea-water conductivity".

Table 1: Parameter's value provided to the initial primary model presumably from design data rules.

"Material 1" polarisation curve p-value	Sea-water conductivity
1.7500	3.0000

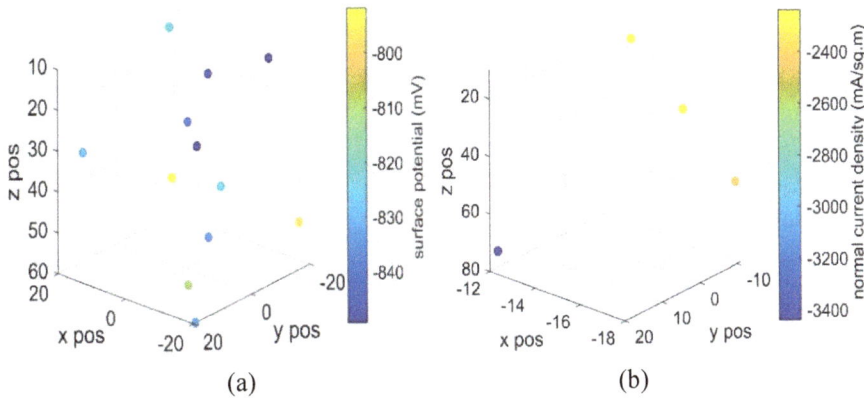

(a)   (b)

Figure 7:   The response surface data. (a) Surface potential; and (b) Current density for the corresponding positions of the validating data in Fig. 6. Obtained after simulation run of initial primary model with parameters value given in Table 1.

4.4.5 Optimisation performance analysis

An initial parametric CP model with provided parameter values as in Table 1 which results in response data as in Fig. 7 on simulation run is considered for optimisation-based parameter updating. Validation data (Section 4.4.2), metric (Section 4.4.3) and optimisation algorithm

(Section 4.4.1) are considered for the optimisation requirement and is performed within the MATLAB based platform.

The discrepancies between the validating data and the model output (initial model and solution model) can be visualised from the Fig. 8, while the optimisation step related insight can be obtained from Table 2, both are generated within the supporting software (MATLAB). The optimisation based parametric search ends when the optimisation-algorithm cannot decrease the objective function further, in the search direction.

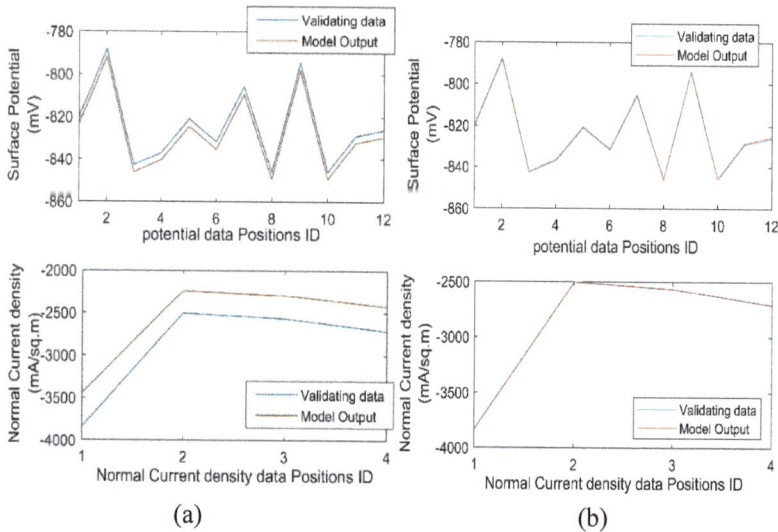

Figure 8: Discrepancies plot between model output and the validating data. (a) Initial primary model; and (b) Solution model after optimisation-based parameter updating.

Table 2: Model parameter stages during the optimisation problem before reaching to a solution.

Iteration	F-count	Material 1 p_value	Sea-water conductivity	Objective function (F(x))
0	3	1.7500	3.0000	5.5168e – 03
1	9	1.8145	3.0443	3.5445e – 03
2	12	2.0516	3.2340	3.5730e – 05
3	15	2.0515	3.2422	2.8556e – 05
4	18	2.0425	3.2655	1.5382e – 05
5	21	2.0229	3.2959	4.5574e – 06
6	24	2.0027	3.3186	1.5651e –06

In the given case, it takes a total of six iterations excluding the initial one to converge to the best set of parameter values. Similarly, counts for repetitive objective function calculation are represented as F-count (Table 2) which is also equivalent to the total data exchange count between server and the simulation, and the total count of simulation runs.

This parameter updating which could take many hours or days when performed manually, is reduced to less than a few hours including all the simulation running time. The importance of the platform is highlighted by the significant reduction of model calibrating time using the automated experimental platform than compared to the manual approach.

5 CONCLUSION

This paper discussed the design of experiments platform for simulation performance enhancement while realising a Digital Twin of a physical asset. The scientific software-based integrated platform forms the basis of automation on enabling virtual replica (Digital Twin) from the pre-available simulator(s), utilising supports from the analytical tool(s).

The case study shows the application of the integration platform to achieve a cathodic protection Digital Twin. Optimisation procedures were undertaken using the approach, and automatic parameter updating of a CP model to calibrate the model with the measured data has been demonstrated.

Overall, the proposed approach combines advantages offered by scientific and commercial software(s) to have a comprehensive tool in reduced time capable to predict the present and future health of a structure. This approach however is generic and can be implemented beyond the corrosion and cathodic protection domain.

Further work will be focused on the automation of the overall customised systematic procedures on enabling DT, besides the optimisation task.

ACKNOWLEDGEMENT

This work has been undertaken as part of a match-funded PhD research between Computational Mechanics International Limited and Bournemouth University, UK.

REFERENCES

[1]   Oliveira, H.L. & Leonel, E.D., Constitutive relation error formalism applied to the solution of inverse problems using the BEM. *Engineering Analysis with Boundary Elements*, **108**, pp. 30–40, 2019.

[2]   Papalambros, P.Y. & Wilde, D.J., *Principles of Optimal Design: Modeling and Computation*, Cambridge University Press, 2000.

[3]   Mitchell, M., Udugama, I., Currie, J. & Yu, W., Software integration for online dynamic simulation applications. *2017 6th International Symposium on Advanced Control of Industrial Processes (AdCONIP)*, IEEE, pp. 360–364, 2017. DOI: 10.1109/ADCONIP.2017.7983807.

[4]   Borodin, V., Bourtembourg, J., Hnaien, F. & Labadie, N., COTS software integration for simulation optimization coupling: Case of ARENA and CPLEX products. *International Journal of Modelling and Simulation*, **39**(3), pp. 178–189, 2019. DOI: 10.1080/02286203.2018.1547814.

[5]   Benaouali, A. & Kachel, S., Multidisciplinary design optimization of aircraft wing using commercial software integration. *Aerospace Science and Technology*, **92**, pp. 766–776, 1 Sep. 2019. DOI: 10.1016/j.ast.2019.06.040.

[6]   Whitfield, R.I., Duffy, A.H., Gatchell, S., Marzi, J. & Wang, W., A collaborative platform for integrating and optimising computational fluid dynamics analysis requests. *Computer-Aided Design*, **44**(3), pp. 224–240, 2012.

[7]   Rasheed, A., San, O. & Kvamsdal, T., Digital twin: Values, challenges and enablers. arXiv preprint arXiv: 1910.01719, 2019.

[8]   Barricelli, B.R., Casiraghi, E. & Fogli, D., A survey on digital twin: Definitions, characteristics, applications, and design implications. *IEEE Access*, **7**(Ml), pp. 167653–167671, 2019. DOI: 10.1109/ACCESS.2019.2953499.

[9]   Abdelmegid, M.A., González, V.A., O'Sullivan, M., Walker, C.G., Poshdar, M. & Ying, F., The roles of conceptual modelling in improving construction simulation studies: A comprehensive review. *Advanced Engineering Informatics*, **46**(Sep.). DOI: 10.1016/j.aei.2020.101175.

[10]   Robinson, S., Arbez, G., Birta, L.G., Tolk, A. & Wagner, G., Conceptual modeling: Definition, purpose and benefits. *2015 Winter Simulation Conference (WSC)*, IEEE, pp. 2812–2826, 2015.

[11]   Brynjarsdóttir, J. & O'Hagan, A., Learning about physical parameters: The importance of model discrepancy. *Inverse Problems*, **30**(11), p. 114007, 2014. DOI: 10.1088/0266-5611/30/11/114007.

[12]   Zienkiewicz, O.C., Taylor, R.L. & Zhu, J.Z., *The Finite Element Method: Its Basis and Fundamentals*, Elsevier, 2005.

[13]   Aliabadi, M.H., *The Boundary Element Method: Applications in Solids and Structures*, vol. 2, John Wiley and Sons, 2002.

[14]   Law, A.M., Kelton, W.D. & Kelton, W.D., *Simulation Modeling and Analysis*, McGraw-Hill: New York, 2000.

[15]   Venter G., Review of optimization techniques. *Encyclopedia of Aerospace Engineering*, 2010.

[16]   Whitley, D., Rana, S., Dzubera, J. & Mathias, K.E., Evaluating evolutionary algorithms. *Artificial Intelligence*, **85**(1–2), pp. 245–276, 1996.

[17]   Boschert, S. & Rosen, R., Digital twin: The simulation aspect. *Mechatronic Futures*, Springer: Cham, pp. 59–74, 2016.

[18]   Wright, L. & Davidson, S., How to tell the difference between a model and a digital twin. *Advanced Modeling and Simulation in Engineering Sciences*, **7**(1), pp. 1–3, 2020. DOI: 10.1186/s40323-020-00147-4.

[19]   Schleich, B., Answer, N., Mathieu, L. & Wartzack, S., Shaping the digital twin for design and production engineering. *CIRP Annals*, **66**(1), pp. 141–144, 2017.

[20]   Inzillo, V., Santamaria, A.F. & Quintana, A.A., Integration of Omnet++ simulator with Matlab for realizing an adaptive beamforming system. *2017 IEEE/ACM 21st International Symposium on Distributed Simulation and Real Time Applications (DS-RT)*, IEEE, pp. 1–2, 2017.

[21]   Sapkota, M.S., Apeh, E., Hadfield, M., Haratian, R., Aey, R. & Baynham, J., An approach for adaptive model performance validation within digital twinning. *CMEM21*, 2021.

[22]   Adey, R.A., *Modelling of Cathodic Protection Systems*, WIT Press: Southampton and Boston, 2006.

[23]   Adey, R., Peratta, C. & Baynham, J., Corrosion data management using 3D visualisation and a digital twin. *NACE International Corrosion Conference Proceedings*, NACE International, pp. 1–13, 2020.

[24]   *MATLAB and Statistics Toolbox Release 2012b*, The MathWorks, Inc.: Natick, MA.

SEA USE MAP: GIS SUPPORTING MARINE AREA'S SUSTAINABLE DEVELOPMENT

MARCO MARCELLI[1,2], FRANCESCO MANFREDI FRATTARELLI[1,2], VIVIANA PIERMATTEI[1,2],
SERGIO SCANU[2], SIMONE BONAMANO[1,2], DANIELE PIAZZOLLA[2] & GIUSEPPE ZAPPALÀ[1,3]
[1]Laboratory of Experimental Oceanology and Marine Ecology (LOSEM, DEB), University of Tuscia, Italy
[2]Ocean Predictions and Applications Division,
Centro Euro-Mediterraneo sui Cambiamenti Climatici (CMCC), Italy
[3]Institute for Biological Resources and Marine Biotechnologies, National Research Council (IRBIM-CNR), Italy

ABSTRACT

The coastal zone is characterised by diversified physical and ecological conditions that allow the multiple use of natural resources. The overlap of very different uses generates conflict inducing habitat depletion and damage to natural systems. To face the problem of "conflict of use" of the coastal areas, we developed an integrated system as a working tool, the Sea Use Map (SUM), aimed to the characterization of the different values and uses of the marine resource, useful to explore further marine uses, such as suitable sites for marine renewable energy production, productive activities, etc. For this reason, the creation of an integrated GIS database, in which the information is conveyed in a geo-referenced system, is the necessary tool to support maritime spatial planning. The SUM of Italy is a key database, in which coastal uses are integrated with environmental data (seabed morphology, waves, currents, fauna, flora, etc.). Further integration between data and simulations of numerical models allows to define the most promising and ecologically acceptable areas for the introduction of new uses in view of the compatible development of the sea and marine resources. This paper presents a pilot application study in the coastal area between Capo Anzio and Tarquinia (Italy, Latium).
Keywords: GIS, marine ecosystem, marine renewable energy, siting, observing system, sea-use.

1 INTRODUCTION

With the Exclusive Economic Zone (EEZ) concept consolidation in the international jurisprudence, the continental shelf will be considered as a national boundary extension. In Italy, with more than 7,000 km of densely populated coasts, the continental shelf is always more affected by fundamental anthropogenic activities. The shelf area is characterized by the extraction of oil and of materials for industries of pottery, metallurgy, and glass; here fishery, mariculture, tourism are practiced; some areas are dedicated to marine biodiversity and environmental protection and to historical heritage protection [1].

In marine coastal areas a conflict among the different uses (both local and remote) occurs [2], [3], therefore it is necessary to minimize reciprocal impacts of the uses preserving at the same time all the structures and ecological processes. Indeed, ecological processes occurring in coastal areas are so important that their contribution to the value of the "natural capital" [4]–[6] is often higher than the terrestrial ones. Natural Capital evaluation is emerging as a fundamental tool to support the management of natural resources, representing the scientific basis for actions needed to enhance their conservation and sustainable use. Indeed, the achievement of the compatibility among their multiple uses, often in conflict in coastal areas, is a priority to avoid the increasing undesirable effects which threat both ecosystems and human health and well-being. Furthermore, the Millennium Ecosystem Assessment [1] assessed the consequences of ecosystems change for human well-being and, in particular, the analysis method has been centered on the linkages between "ecosystem services" (i.e., the benefits that people obtain from ecosystems) and human well-being.

WIT Transactions on Engineering Sciences, Vol 130, © 2021 WIT Press
www.witpress.com, ISSN 1743-3533 (on-line)
doi:10.2495/CMEM210021

Within the Blue Growth strategy, the protection of marine ecosystems is considered a priority for the sustainable growth of marine and maritime sectors. To face this issue, the European MSP and MSFD directives [7], [8] strongly promote the adoption of an ecosystem-based approach; particular attention is paid to support monitoring networks that use Long-Term Ecological Research (L-TER) observations and integrate multi-disciplinary data sets and to define criteria for maritime space management. For all these reasons, there is the need to develop a methodology to analyze marine ecosystems properties in relationship within sea uses and climate change scenarios. An interdisciplinary approach including oceanography, ecology, geology, biology, meteorology is the basis of an integrated knowledge [9] able to support the decision through GIS and scenario's simulation.

This study aims to minimize the conflict among the different uses, creating a working tool able to support the decision, in order to select the best use for a specific marine area.

To face this necessity we developed a multi-layer GIS of coastal marine areas (Sea Use Map, SUM, Fig. 1), as a part of an integrated observing system, supported by a permanent monitoring system and able to interface a hierarchy of mathematical models, both for their validation and for the simulation of scenarios. This system takes into account both the different sea uses and the value of marine ecosystems, calculated on the basis of services and benefits produced by the different biocoenoses [2], [3].

The first prototype has been applied on the northern Latium coasts [10] in order to support the sustainable development of the Civitavecchia harbour.

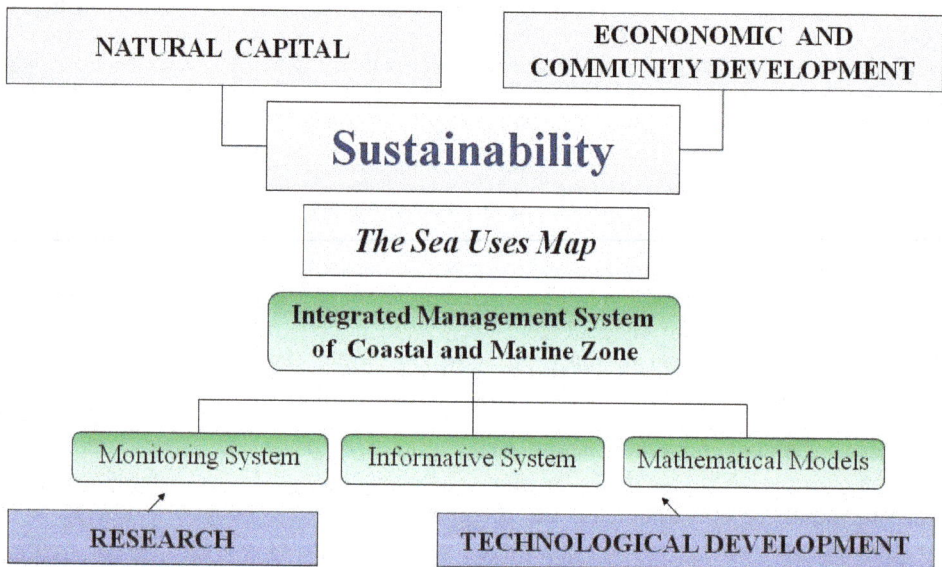

Figure 1: The Sea Use Map represents the connection tool between sustainable development and Natural Capital conservation on one hand and the technical tools to support the decision-making system on the other. The three modules (monitoring system, information system and numerical models) provide data and scenarios that overlap the uses allowing to guide the decision. Research and technological development fosters the three modules which remain up-to-date and provide increasingly reliable information to support decision making.

This tool supports a sustainable management of marine coastal areas, basically offering ecosystem benefits evaluation and pollution impacts minimization (Fig. 2), such as the selection of the best sites for the introduction of new uses or the identification of the coastal areas subjected to potential impacts. In synthesis, SUM is an ecosystem-oriented cartographic tool specifically designed to integrate the Civitavecchia Coastal Environment Monitoring System (C-CEMS) [10], consisting of: coastland maritime space uses, geomorphological and ecological features, mathematical modelling results, forecast scenarios, satellite observations, and *in situ* data from meteorological stations, wave buoys and x-band radar, seawater fixed stations based on low cost technological developments providing physical and biological data, and making periodic surveys to characterize benthic biocoenoses and sediments.

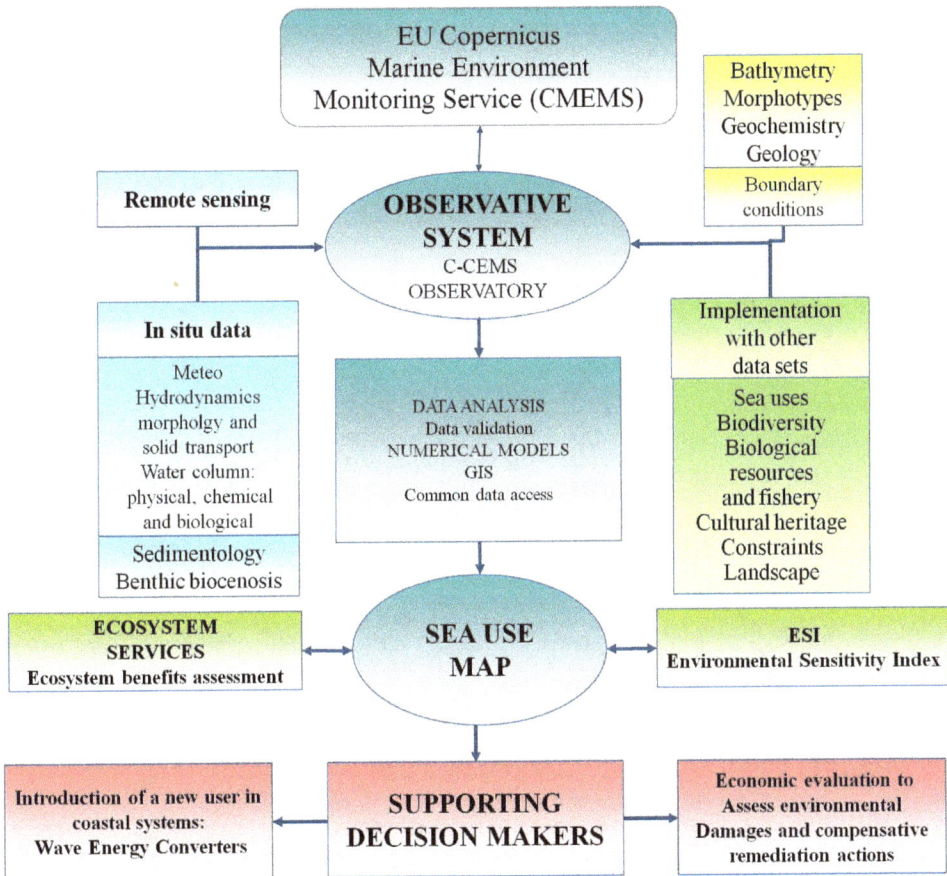

Figure 2:   Diagram of components and goals of SUM as a scheme of an Integrated Observing System to be applied in the coastal and marine zones.

Our study focuses on the central and northern Latium (Italy) coastal area, characterized by high variability of marine and coastal environments, hosting important biodiversity hotspots (*Posidonia oceanica*, Reefs, protected species as *Pinna nobilis* and *Corallium rubrum*), historical heritage, and affected by the presence of the second biggest Italian river basin (Tiber River), of one of the most important port for cruise traffic in the Tyrrhenian Sea (The Port of Civitavecchia) [11], relevant industrial infrastructures, touristic features, and other minor ports (Anzio, Fiumicino, Santa Marinella).

The chosen approach aims to overcome the conflict of uses by integrating, in a system of tested numerical models and synoptic representations, information from field measurements and remote images. The integration of remote sensing data with field data has a critical importance because the latter represents the sea truth of the satellite synopsis (Fig. 3). This data integration allows a detailed analysis of the evolution in space and time of the effects due to the interactions between anthropogenic activities and natural ocean processes. The calibrated satellite data and the *in situ* data can in turn be assimilated into mathematical models through which it will be possible to obtain the representation of various scenarios. The present approach is therefore based on an overall strategy of detection, data processing and modeling, which will contribute to the evaluation of the space-time variation of the studied processes.

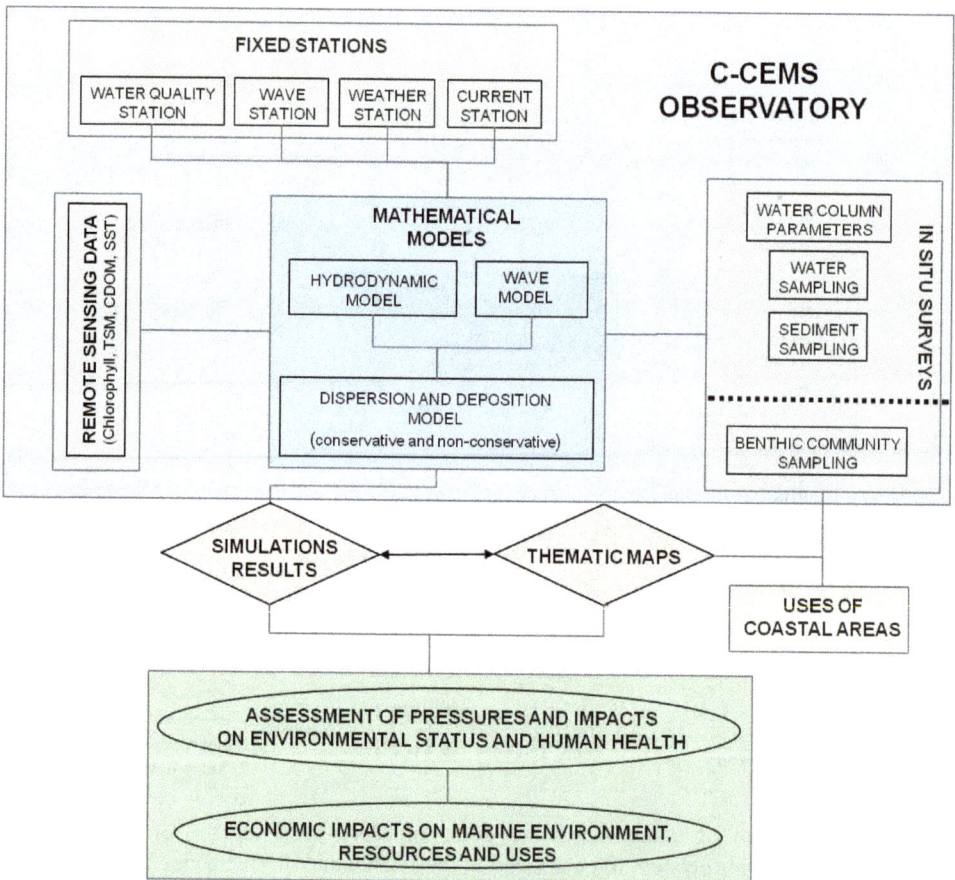

Figure 3:   Main components of the Integrated Observing System "The Civitavecchia Coastal Environment Monitoring System" (C-CEMS) [applied to Civitavecchia port 10].

The SUM system is the result of an integration between environmental data collected by monitoring activities, socio-economic data coming from coastal zone uses, simulations by mathematical models and geographic information platform. It includes information needed for the planning of some activities connected to the coastal zone. This system can be divided into three main components (Fig. 1):

- Monitoring system;
- Informative system; and
- Mathematical models.

Within the environmental monitoring component can be detected different modules such as:

- coastal zone influence: this module inserts in the management system information about coastal morphotypes, water and solid outflows which characterize near shore marine zones, geology and geochemistry and about the land use in hinterland zones;
- sea bottom: this module integrates detailed information about morphological and geological characteristics of the seabed and information on the distribution and abundance of marine biota;
- marine weather: the marine weather data are useful to know the main forcing conditions on the sea interface such as wind, waves, tides and currents;
- water column: this module analyses physical, chemical and biological variables measured along the water column; and
- uses of marine and coastal areas: this module is crucial for planning coastal activities, to monitor and forecast their effects.

Management system's pillars are represented by previous data which will be gradually enriched with the new monitoring data collected.

In conclusion, this integrated system will be applied for the following purposes:

- management and data visualization;
- decision making;
- integrated database to support Environmental Impacts Study and to support authorization procedures for infrastructures realization;
- ante and post-operam studies;
- the simulation of different scenarios through mathematical models; and
- guidelines for monitoring plans.

1.1 Description of the components

The monitoring component is divided into different interdependent and flexible components, at different time and space scale. It is also included in MonGOOS system (see Fig. 4), and it can be easily and quickly integrated and/or modified on the basis of numerical modelling. Monitoring includes:

- experimental activities aimed at validating numerical models. In particular: oceanographic measurements of waves, currents, characteristics of the water column and sedimentological and solid transport measurements (including river inputs);

- analysis of the morphodynamic sector; study of the liquid and solid flows of the Tiber river and the main tributaries of water;
- study of the variation of the shoreline in the short and long term through historical analysis of the shoreline and reconstruction of the submerged beach profiles (starting from 1990) and subsequent modifications using GIS; preparation and simulations through the model for the analysis of the shoreline evolution in the short and long term;
- preparation of the operating model and validation activities;
- characterization of the biotic compartment and of the chemical characteristics of the sediments;
- fixed measurement platforms (Acoustic Doppler Current Profiler – ADCP; wave buoys);
- autonomous measurement platforms in order to have more information, optimizing the sampling plan without increasing costs; and
- remote sensing from satellite (temperature, altimetry, chlorophyll a, Chromophoric Dissolved Organic Matter – CDOM, turbidity).

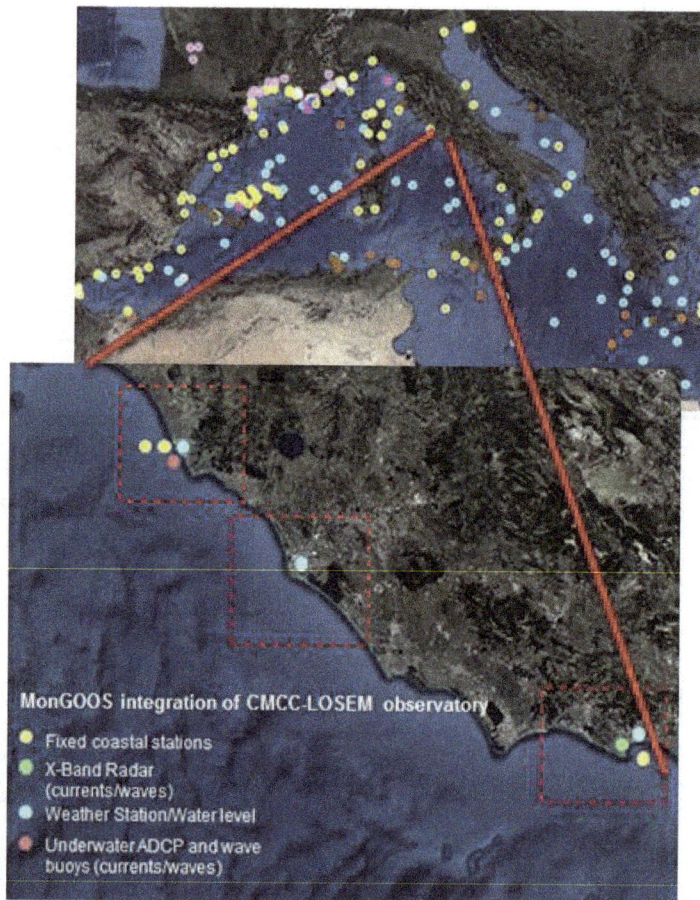

Figure 4:   Fixed measurement stations and their location with respect to the MonGOOS system (*http://www.mongoos.eu/*).

Below are reported some GIS representations and examples of use of the SUM in the coastal area of northern Latium (Italy). The continental component takes into account not only the fluvial contributions but also coastal morphotypes and land cover (Fig. 5). These

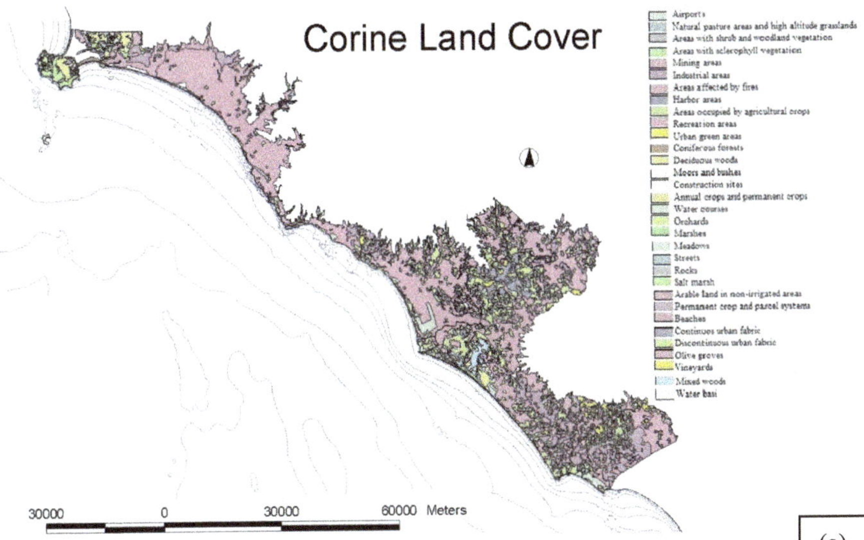

Corine Land Cover

Legend:
- Airports
- Natural pasture areas and high altitude grasslands
- Areas with shrub and woodland vegetation
- Areas with sclerophyll vegetation
- Mining areas
- Industrial areas
- Areas affected by fires
- Harbor areas
- Areas occupied by agricultural crops
- Recreation areas
- Urban green areas
- Coniferous forests
- Deciduous woods
- Moors and bushes
- Construction sites
- Annual crops and permanent crops
- Water courses
- Orchards
- Marshes
- Meadows
- Streets
- Rocks
- Salt marsh
- Arable land in non-irrigated areas
- Permanent crop and parcel systems
- Beaches
- Continuous urban fabric
- Discontinuous urban fabric
- Olive groves
- Vineyards
- Mixed woods
- Water basin

30000 0 30000 60000 Meters

(a)

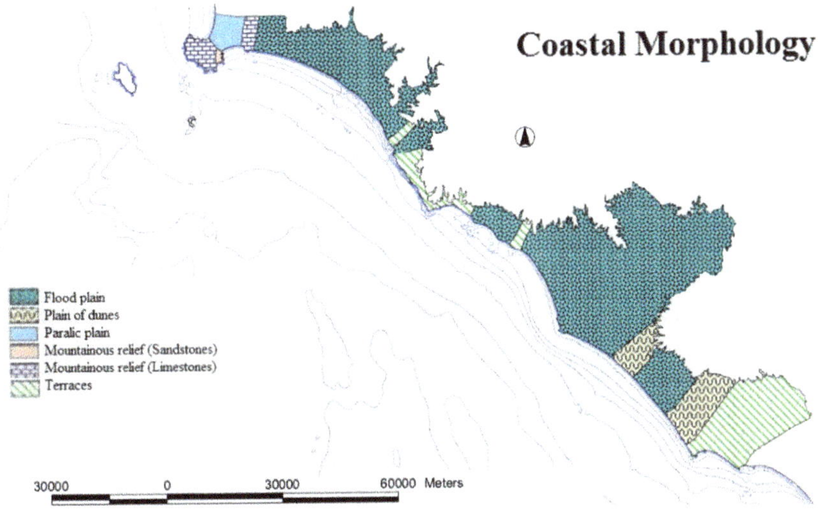

Coastal Morphology

Legend:
- Flood plain
- Plain of dunes
- Paralic plain
- Mountainous relief (Sandstones)
- Mountainous relief (Limestones)
- Terraces

30000 0 30000 60000 Meters

(b)

Figure 5: (a) Representation of the territory using the method of the European project Corine Land Cover for the preparation of the land use map. In this project, each element of the territory (pixel of 100 m × 100 m) is associated with a unique use based on a 3-level classification. Furthermore, the analysis of the territory is performed through remote sensing images; and (b) Representation of coastal morphotypes, based on the classification reported by Brondi et al. [14].

are useful for computing different climatic parameters and also to represent the cultural, recreational and social benefits useful to calculate the Cultural Ecosystem Services (CES).

The marine areas are represented by integrating a lot of information, from instrumental mapping of the seabed to the precise data coming from monitoring activities (e.g., bio-optic characterization of the water column, characterization of the benthic biocoenoses) (Figs 6 and 7). The integrated tool SUM therefore allows us to weigh the uses of the sea and their effects on the ecosystems. For example, Fig. 8 shows the use of SUM in relation to fishing activity. For fishing data, a three-digit code is proposed, each of them relates to a kind of fishing: the first on the left represents trawling, the one in the centre represents small-scale fishing and the one on the right indicates shellfish fishing. Each digit assumes the value 0 if that activity is not practiced, the value 1 in the opposite case and the values 2, 3 … 9 depending on the type of fishing performed by the 1, 2 … fleet (Fig. 8).

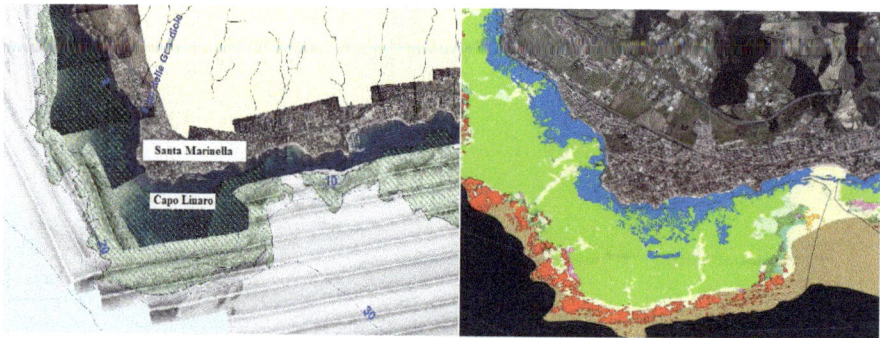

Figure 6:   (Left) GIS representation of *Posidonia oceanica* meadow integrating several survey methods and (right), GIS classification of biocoenoses.

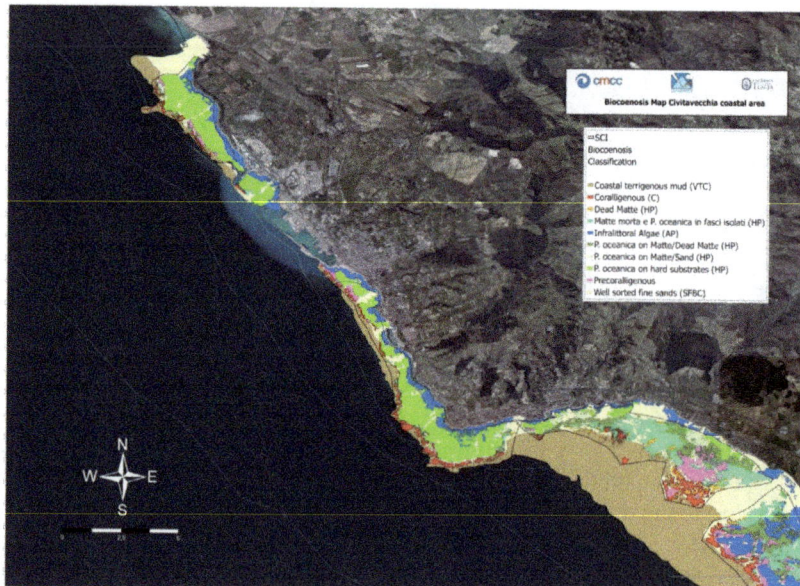

Figure 7: GIS representation of benthic biocoenoses in the northern Latium coastal area.

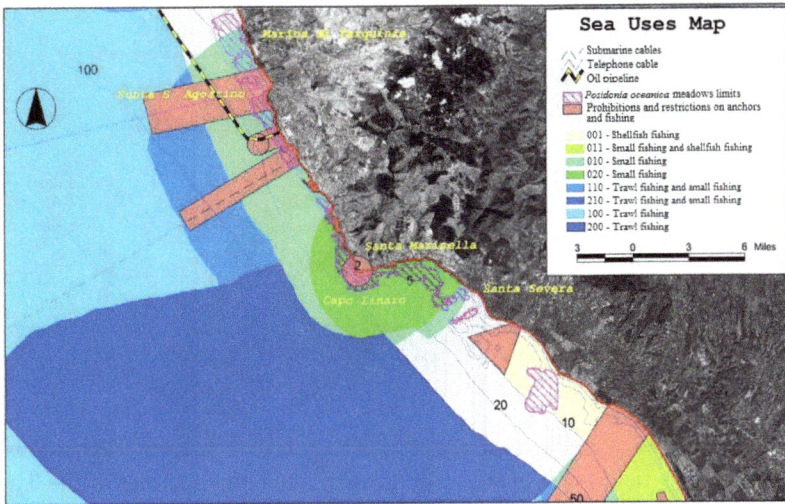

Figure 8:   Fishing activity analysed using SUM in the coastal area of Civitavecchia (Northern Latium, Italy).

1.2 Characteristics of the numerical models

Current velocity and wave propagation are obtained by numerical models. Here we present preliminary results obtained by simulations between S. Agostino and Capo Linaro. In particular, current velocity field is simulated by ADvanced CIRCulation model (ADCIRC [12]) and wave field is simulated by the STeady-state spectral WAVE model (STWAVE [13]). ADCIRC model solves the motion equations for a moving fluid on a rotating earth. Those equations are formulated by using the traditional hydrostatic pressure and Boussinesq approximations; they are discretized in space by using the finite elements method, and in time by using the finite differences method. The computational domain covers the coastal area for about 150 km. Because small errors could occur near the boundaries, the study area was located far from the model edges. The fine element mesh shows a resolution ranging from 30 m in the areas surrounding Civitavecchia port to about 5 km near the off-shore boundary positioned about 30 km from the coast.

The current velocities, calculated by the model, are obtained combining the general circulation current of Tyrrhenian Sea (as boundary condition), and the south-east wind (Scirocco – 135°N) with an intensity of 10 m/s.

The Scirocco wind increases the current velocity to the south (0.5 m/s), at Capo Linaro, and to the north (0.3 m/s) near S. Agostino (Fig. 9) while in Civitavecchia harbour the velocities are very slow.

The wave field is obtained by STWAVE model, which is a finite-differences model solving the wave dispersion equation and take into account of refraction, shoaling, diffraction and radiation stress. The computational domain extends from S. Agostino in the north to Capo Linaro in the south. The offshore boundary is located at 70 m depth. The model grid presents a constant resolution with the element size of 50 m. The diffraction intensity is equal to 1 and the bed friction is assigned to 0.005 and assumed spatially constant in the entire domain. The wave field is obtained by wave spectrum calculated with wave height of 3 m and direction of 225°N.

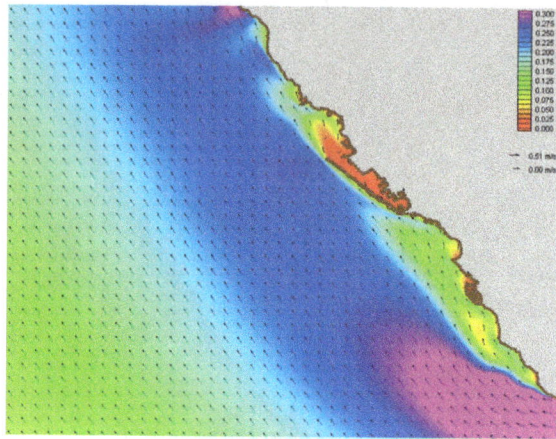

Figure 9: Velocity field (m/s) obtained by general circulation current of the Tyrrhenian Sea and by Scirocco wind (135°N). Purple areas represent high velocity while those in red represent low velocity.

Fig. 10 left shows the wave propagation towards the coast and the areas where wave energy is high (dark blue zone). Fig. 10 right represents the velocity field obtained by radiation stress calculated from a wave height of 3 m and a direction of 225°. The results of the simulation show an increase in velocity (0.4 m/s) in the northern part of the study area.

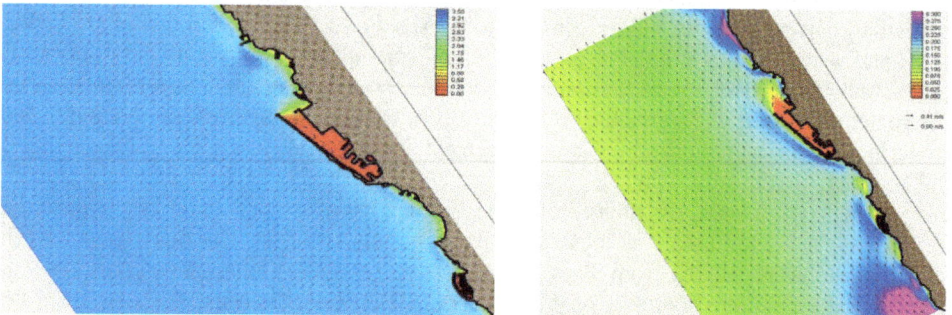

Figure 10: (Left) Wave field obtained from STWAVE simulation. Blue areas represent high wave elevations while those in red represent low wave elevations. (Right) Velocity field obtained by radiation stress gradient. Purple areas represent high velocity while those in red represent low velocity.

2 DISCUSSION AND CONCLUSIONS

Coasts are extremely sensitive areas, characterized by the presence of multiple human activities due to the simultaneous presence of physical and ecological conditions advantageous for the multiple use of natural resources. To address this problem, a strategy based on the best evaluation of the conditions is needed, which determine the choice of the best use of a specific area. This strategy must be based on the best evaluation of the conditions, which determine the choice of the best use of a specific area. The choice takes into account all the components of the system and the effects that come from the choice.

Some problems can arise and the perception of these problems can be:

- oriented by a resource or by a specific status of the resource (regardless of the use of the resource, like littoral dynamic, rare species, whole ecosystems) and
- uses oriented: *threshold values* of some descriptors can damage or preclude a specific use (the value can be or not arranged by normative references, e.g., "no bathing" areas).

Both the perceptions, uses and resources, must be considered in the planning action.

To cope with the described problem, we have developed the Sea-Use-Map (SUM), an integrated system and work tool. The SUM is the only instrument which allows both to analyse all the components of a system and to weigh uses conflict and effects.

Some problems remain: the analyses of the stored information and the choice of the analytical instruments to be integrated in the SUM, in order to provide numerical index for the decision support. Until now the choice is oriented to the integration of decision making models, which seem to be pointed to this approach.

REFERENCES

[1] Millennium Ecosystem Assessment (MA), Overview of the millennium ecosystem assessment. www.millenniumassessment.org/en/About.html. Accessed on: 15 Oct. 2018.
[2] Marcelli, M., Scanu, S., Frattarelli, F.M., Mancini, E. & Carli, F.M., A benthic zonation system as a fundamental tool for natural capital assessment in a marine environment: A case study in the Northern Tyrrhenian Sea, Italy. *Sustainability*, 10(10), p. 3786, 2018.
[3] Marcelli, M., L'importanza dello studio della dinamica litoranea nella progettazione portuale. *Chiron*, 1, pp. 10–12, 2003.
[4] Costanza, R. et al., The value of the world's ecosystem services and natural capital. *Nature*, 387(6630), pp. 253–260, 1997.
[5] Costanza, R., Pérez-Maqueo, O., Martinez, M.L., Sutton, P., Anderson, S.J. & Mulder, K., The value of coastal wetlands for hurricane protection. *Ambio*, 37(4), pp. 241–248, 2008.
[6] Costanza, R. et al., Changes in the global value of ecosystem services. *Global Environmental Change*, 26, pp. 152–158, 2014.
[7] EC Directive 2014/89/EU of the European Parliament and of the Council of 23 July 2014 establishing a framework for maritime spatial planning, 2014.
[8] EC Directive 2008/56, *Official Journal of the European Union*, 164(25 Jun. 2008), pp. 19–40, 2008.
[9] Brondi, A., Di Maio, A. & Marcelli, M., Principi logici per il monitoraggio dell'ambiente costiero. *Geologia dell'Ambiente*, 2, pp. 19–35, 2008.
[10] Bonamano, S. et al., The Civitavecchia Coastal Environment Monitoring System (C-CEMS): A new tool to analyze the conflicts between coastal pressures and sensitivity areas. *Ocean Science*, 12(1), pp. 87–100, 2016.
[11] Piazzolla, D. et al., Microlitterpollution in coastal sediments of the northernTyrrhenian Sea, Italy: Microplastics and fly-ash occurrence and distribution. *Estuarine, Coastal and Shelf Science*, 241, p. 106819, 2020.

[12] Luettich, R.A. & Westerink, J.J., AADvance CIRCulation model for oceanic, coastal and estuarine waters. Manual, 2000.

[13] Smith, J.M., Sherlock, A.R. & Resio, D.T., STWAVE: Steady-state spectral wave model. *User's Manual for STWAVE*, Version 3.0, ERDC/CHL SR-01-1, Coastal and Hydraulics Laboratory, 2001.

[14] Brondi, A., Benvegnù, F., Ferretti, O. & Anselmi, B., Classificazione geomorfologica delle coste italiane come base per l'impostazione di studi sulla contaminazione marina. *Proceedings of the Italian Association of Oceanology and Limnology (AIOL)*, Sorrento, 1982.

ON SOME NUMERICAL TECHNIQUES FOR STRESS-CONSTRAINED CONTINUA

MASSIMILIANO LUCCHESI, BARBARA PINTUCCHI & NICOLA ZANI
Department of Civil and Environmental Engineering (DICeA), University of Florence, Italy

ABSTRACT

A numerical method is described for the semi-explicit solution of the constitutive equation of an orthotropic non-linear elastic material with various constraints on the stress. Each constraint forces the stress to belong to a closed and convex cone in the space of second order symmetric tensors, and via the choice of appropriate values cones vertex is possible to assign different strength characteristic in different directions. The constitutive equation thus formulated allows to model masonry with very general textures. However application to problems of equilibrium and evolution requires some numerical expedients which are described in detail in this paper. The equation, implemented in the MADY finite element code, has been used to examine how the masonry panel strength changes as a function of traction directions. The results have been compared with the analougous obtained modelling the masonry at the micro scale.

Keywords: orthotropic materials, masonry panels.

1 INTRODUCTION

A constitutive model for masonry materials requires accounting for its poor tensile strength. This need led to the study of the masonry-like material [1], which in its original formulation imposes the constraint on the stress T of being semi-defined negative. The equation was then generalized by also imposing a limit on compressive and shear strength [2], [3]. Each of these constraints requires T to belong to a specific closed and convex cone, whose intersection is the stress range $\mathcal{K}$. Having assigned a strain tensor E and the symmetric and positive-definite tensor $\mathbb{C}$ of the elastic moduli, the stress T is obtained by projecting $\mathbb{C}E$ onto $\mathcal{K}$, with respect to a suitable scalar product. In other words, it is required that the inelastic strain $E^a = E - \mathbb{C}^{-1}T$ belongs to the normal cone of $\mathcal{K}$ in T. The existence and uniqueness of the projection follows by the minimum norm theorem and the result is a hyperelastic material which has been defined normal-elastic material.

In order to apply this constitutive equation to engineering problems, it is necessary to resort to numerical analysis. This in turn required that the constitutive equation be explicitly solved and the derivative of the stress with respect to the strain explicitly calculated [4]. With the constraints on the stress that were taken into account, $\mathcal{K}$ resulted to be a spherical set, i.e. such that $\mathcal{K} = Q\mathcal{K}Q^T$, for each rotation Q, and moreover $\mathbb{C}$ was hypothesized to be an isotropic tensor. The first of these properties implies the coaxiality between the stress and the inelastic strain, and the second (if the first is guaranteed) implies the coaxiality between the stress and the strain. The fact that T, E and E^a are all coaxial tensors allowed to solve the constitutive equation in their common characteristic space, simplifying the problem. This constitutive equation has been applied to the study of numerous monuments and may be effective for certain buildings where the texture and the properties of the masonry cannot be easily evaluable. On the other hand, for different applications a model that takes into account the different properties of the material in different directions is certainly more realistic [5], [6]. For this reason in [7] the equation has been generalized allowing the tensor

WIT Transactions on Engineering Sciences, Vol 130, © 2021 WIT Press
www.witpress.com, ISSN 1743-3533 (on-line)
doi:10.2495/CMEM210031

of the elastic moduli to be orthotropic. Consequently, the coaxiality between T and E has been lost, but that between T and E^a is maintained.

In [8], the material has been supposed to have different strength characteristics in different directions, a circumstance that can often occur due to the constructive techniques and to material's damage process. In such a case, the stress range $\mathcal{K}$ is made by the intersection of several closed and convex cones each of which has as its vertex a tensor that is not necessarily spherical. Thus, the coaxiality between T and E^a is lost. Despite the greater complexity that must be tackled, the strain space is always naturally divided into a finite number of regions and E belong to one of these determines the solution of the constitutive equation. Thanks to this property, albeit with some expedient, it is possible to obtain an "almost explicit" solution of the constitutive equation. However, this generalization requires new numerical procedures, mainly because it is not possible to determine a priori the region to which the deformation belongs, and therefore it is necessary to proceed by trials and errors.

In this paper, the numerical techniques that allow to apply this constitutive equation to equilibrium and evolution problems is described in detail. Then the equation, implemented in the finite element code MADY, is used to solve same test case. In particular, a panel with a typical texture is considered and analyzed at the microscale by subjecting it to a tensile force that is applied in five different directions and increased until the collapse. Subsequently, a "homogeneous" panel with a tensile strength deduced from the previous test is subjected to the same loads. It is evidenced how the texture can be well modelled by a suitable choice of the vertices of $\mathcal{K}$.

2 NOTATIONS AND BACKGROUND

Let Sym be the space of the symmetric second order tensor with the inner product $A \cdot B = \mathrm{tr}(AB)$, and the corresponding Euclidean norm $\| \ \|$. The energetic inner product $(A, B)_E = A \cdot \mathbb{C}^{-1} B$ and the corresponding norm denoted by $\| \ \|_E$ will also be considered. Moreover, if $\mathcal{K} \subset$ Sym is the non empty, closed and convex set made by all the admissible stresses, for each $T \in \partial\mathcal{K}$ (the boundary of $\mathcal{K}$), the normal cone of $\mathcal{K}$ at E is denoted by $\mathcal{N}(\mathcal{K}, T)$.

Assigned a strain tensor E, there exists only one element $T \in \mathcal{K}$ having the minimum energy distance from $\mathbb{C}E$, i.e. such that

$$\|\mathbb{C}E - T\|_E = \min_{S \in \mathcal{K}} \|\mathbb{C}E - S\|_E. \tag{1}$$

This is equivalent to saying that T is the projection of $\mathbb{C}E$ onto $\mathcal{K}$, with respect to the energetic norm, i.e. $E - \mathbb{C}^{-1}T$ belongs to $\mathcal{N}(\mathcal{K}, T)$.

Relation $T = \hat{T}(E)$ defines the constitutive equation of a non linear hyperelastic material that is termed normal elastic material [1]. Once set

$$E^e = \mathbb{C}^{-1}T, \quad E^a = E - \mathbb{C}^{-1}T, \tag{2}$$

it turns out $E = E^e + E^a$; E^e and E^a are said, *elastic* and *inelastic* part of the deformation, respectively In particular, if T is a regular point of $\partial\mathcal{K}$ and $N(T)$ is the corresponding unit outward normal, then

$$E^a = \alpha N(T), \quad \alpha \geq 0, \tag{3}$$

holds. Note that if $\mathcal{K}$ is a spherical tensor then $N(T)$ is an isotropic function of T and then T and E^a have the same characteristic space so that $TE^a = E^aT$.

From eqn (2)

$$\mathbb{C}E - T = \alpha\mathbb{C}N(T)$$

follows and then, with a few substitutions,

$$T = \mathbb{C}E - \frac{(\mathbb{C}E - T) \cdot N}{N \cdot \mathbb{C}N} \mathbb{C}N = \mathbb{C}E - \frac{(\mathbb{C}N \otimes N)}{N \cdot \mathbb{C}N} (\mathbb{C}E - T) \tag{4}$$

is obtained.

Let be $\mathcal{E} = \mathbb{C}^{-1}(\mathcal{K})$; if $T = \hat{T}(E) \in \partial\mathcal{K}$ then $E \in \partial\mathcal{E}$ and, if T is a regular point of $\partial\mathcal{K}$ also E is a regular point of $\partial\mathcal{E}$, with outward unit normal $M(E) = \mathbb{C}N(\hat{T}(E))/\|\mathbb{C}N(\hat{T}(E))\|$.

A convex cone with vertex V is a non-empty, closed set $\mathcal{C}$ containing the origin 0 of Sym, such that

$$V + a(S - V) + b(T - V) \in \mathcal{C}, \text{ for each } S, T \in \mathcal{C} \text{ and for each } a \geq 0 \text{ and } b \geq 0. \tag{5}$$

Proceeding in a similar way to what done in [9], it can be verified that for this set the normal cone $\mathcal{N}(\mathcal{C}, T)$ is made up of all the elements $A \in$ Sym such that

$$\text{(i)} \quad (T - V) \cdot A = 0, \quad \text{(ii)} \quad (S - V) \cdot A \leq 0, \text{ for each } S \in \mathcal{C}. \tag{6}$$

Let $T \in \partial\mathcal{C}$ be a regular point and $N(T)$ the corresponding outward unit normal. Then, in view of eqn $(6)_1$ it holds

$$(T - V) \cdot N(T) = 0. \tag{7}$$

This implies

$$(\mathbb{C}E - T) \cdot N(T) = (\mathbb{C}E - T + T - V) \cdot N(T) = (\mathbb{C}E - V) \cdot N(T),$$

and then from eqn (4) it follows

$$T = \mathbb{C}E - \frac{(\mathbb{C}N \otimes N)}{N \cdot \mathbb{C}N} (\mathbb{C}E - V), \tag{8}$$

which is more convenient relation than eqn (4) because in applications V is known.

3 MATERIALS WITH LIMITED TENSILE, COMPRESSIVE AND SHEAR STRENGTH

The Let Sym$^-$ and Sym$^+$ the subsets of Sym consisting of the negative and positive semi definite tensors, respectively, and let T_t, T_c and T_s be positive semi definite tensors. A material is said to have limited tensile, compressive and shear strength if the stress is constrained to belonging to the closed and convex set $\mathcal{K}$ which is made by the intersection of the three cones

$$\mathcal{T} = \{T \in \text{Sym} : \text{tr}(T - T_t) \leq 0, \quad \det(T - T_t) \geq 0\}, \tag{9}$$

$$\mathcal{C} = \{T \in \text{Sym} : \text{tr}(T + T_c) \geq 0, \quad \det(T + T_c) \geq 0\} \tag{10}$$

and

$$\mathcal{S} = \{T \in \text{Sym} : \|T - T_s\|^2 - \frac{1 + \sin^2\phi}{2}(\text{tr}(T - T_s))^2 \leq 0, \text{tr}(T - T_s) \leq 0\} \tag{11}$$

with vertex T_t, T_c and T_s, respectively, and whose boundary is made up, in addition to $\{T_t\}$, $\{T_c\}$ and T_s, of the sets

$$\{T \in \text{Sym} : \text{tr}(T - T_t) < 0, \quad \det(T - T_t) = 0\}, \tag{12}$$

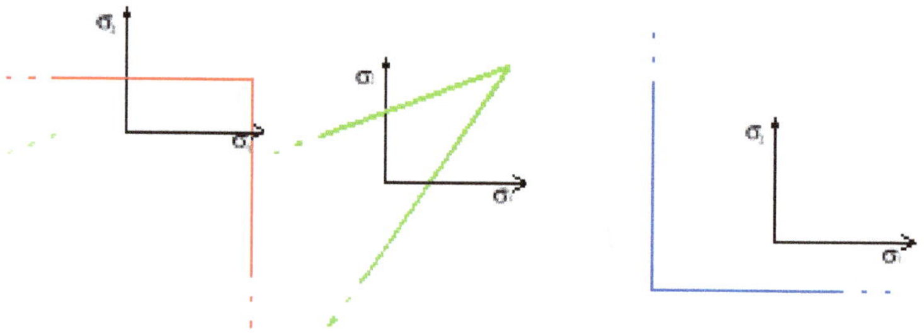

Figure 1: The stress constraints.

$$\{\boldsymbol{T} \in \text{Sym} : \text{tr}(\boldsymbol{T} + \boldsymbol{T}_c) > 0, \quad \det(\boldsymbol{T} + \boldsymbol{T}_c) = 0\} \tag{13}$$

and

$$\{\boldsymbol{T} \in \text{Sym} : 2\|\boldsymbol{T} - \boldsymbol{T}_s\|^2 - (1 + \sin^2\phi)\text{tr}(\boldsymbol{T} - \boldsymbol{T}_s)^2 = 0, \quad \text{tr}(\boldsymbol{T} - \boldsymbol{T}_s) < 0\}. \tag{14}$$

The three cones are shown in Fig. 1, with different colors because they are represented in the characteristic spaces of $\boldsymbol{T} - \boldsymbol{T}_t$, $\boldsymbol{T} + \boldsymbol{T}_c$ and $\boldsymbol{T} - \boldsymbol{T}_s$, which in general are different from each other. It comes down to the usual case when $\boldsymbol{T}_t$, $\boldsymbol{T}_c$ and $\boldsymbol{T}_s$ are spherical tensors, and the coaxiality of $\boldsymbol{T}, \boldsymbol{T}_t, \boldsymbol{T}_c$ and $\boldsymbol{T}_s$ allows to represent $\mathcal{K}$ in the characteristic space of $\boldsymbol{T}$, as shown in Fig. 2(a).

4 DETERMINATION OF THE STRESS

Let the orthonormal vectors $\boldsymbol{e}_1$ and $\boldsymbol{e}_2$ define the symmetry directions of a plane orthotropic body [10] which are assumed as refence system. The elasticity tensor is

$$\mathbb{C} = \begin{pmatrix} C_{1111} & C_{1122} & 0 \\ C_{1112} & C_{2222} & 0 \\ 0 & 0 & C_{2323} \end{pmatrix},$$

where

$$C_{1111} = \frac{E_{11}}{1 - \nu_{12}\nu_{21}}, \quad C_{1122} = \frac{\nu_{12}E_{22}}{1 - \nu_{12}\nu_{21}} = \frac{\nu_{21}E_{11}}{1 - \nu_{12}\nu_{21}},$$

$$C_{2222} = \frac{E_{22}}{1 - \nu_{12}\nu_{21}}, \quad C_{2323} = 2G,$$

with E_{11}, E_{22} and ν_{12}, ν_{21} the corresponding Young and Poisson moduli, respectively.

By denoting

$$E_{22} = \beta E_{11}, \quad \nu_{21} = \beta\nu_{12}, \quad 2G = \frac{\phi E_{11}}{1 - \beta\nu_{12}^2}$$

and writing E for E_{11} and ν for ν_{12}, it is obtained

$$\mathbb{C} = \frac{E}{1 - \beta\nu^2} \begin{pmatrix} 1 & \beta\nu & 0 \\ \beta\nu & \beta & 0 \\ 0 & 0 & \phi \end{pmatrix} \tag{15}$$

and

$$\mathbb{C}^{-1} = \frac{1}{E} \begin{pmatrix} 1 & -\nu & 0 \\ -\nu & 1/\beta & 0 \\ 0 & 0 & (1 - \beta\nu^2)/\phi \end{pmatrix}. \tag{16}$$

In the particular case of $\beta = 1$ and $\phi = 1 - \nu$, the material is isotropic. Let

$$\boldsymbol{E} = \epsilon_{11}\boldsymbol{e}_1 \otimes \boldsymbol{e}_1 + \epsilon_{22}\boldsymbol{e}_2 \otimes \boldsymbol{e}_2 + \epsilon_{12}(\boldsymbol{e}_1 \otimes \boldsymbol{e}_2 + \boldsymbol{e}_2 \otimes \boldsymbol{e}_1)$$

be the assigned strain tensor so that

$$\mathbb{C}\boldsymbol{E} = \frac{E}{1 - \beta\nu^2}\bigg((\epsilon_{11} + \beta\nu\epsilon_{22})\boldsymbol{e}_1 \otimes \boldsymbol{e}_1 + \beta(\epsilon_{22} + \nu\epsilon_{11})\boldsymbol{e}_2 \otimes \boldsymbol{e}_2$$
$$+ \phi\epsilon_{12}(\boldsymbol{e}_1 \otimes \boldsymbol{e}_2 + \boldsymbol{e}_2 \otimes \boldsymbol{e}_1)\bigg). \tag{17}$$

In order to solve the constitutive equation a semi explicit procedure has been developed. All constraints are expressed in the reference system.

Let $\mathcal{T}$ be the traction cone with vertex

$$\boldsymbol{T}_t = \sigma_{tx}\mathbf{e}_1 \otimes \mathbf{e}_1 + \sigma_{ty}\mathbf{e}_2 \otimes \mathbf{e}_2 + \tau_t(\mathbf{e}_1 \otimes \mathbf{e}_2 + \mathbf{e}_2 \otimes \mathbf{e}_1). \tag{18}$$

The eigenvalues $(\sigma_{t_1}, \sigma_{t_2})$ are the values of the tensile strength referred to the corresponding principal directions $(\boldsymbol{e}_{t_1}, \boldsymbol{e}_{t_2})$. In the characteristic space, the cone can be represented as a plane region as shown in Fig. 1(a).

Analougously, the shear cone $\mathcal{S}$ has vertex

$$\boldsymbol{T}_s = \frac{\tau_{0x}}{m}\boldsymbol{e}_1 \otimes \boldsymbol{e}_1 + \frac{\tau_{0y}}{m}\boldsymbol{e}_2 \otimes \boldsymbol{e}_2 + \frac{\tau_s}{m}(\boldsymbol{e}_1 \otimes \boldsymbol{e}_2 + \boldsymbol{e}_2 \otimes \boldsymbol{e}_1). \tag{19}$$

Its eigenvalues $(\frac{\tau_{0_1}}{m}, \frac{\tau_{0_2}}{m})$ are the ratio between coesion and $m = tan\ \phi$, referred to the principal directions $(\boldsymbol{e}_{s_1}, \boldsymbol{e}_{s_2})$. In this characteristic space, such constraint can be represented as a plane region, as shown in Fig. 1(b).

Lastly, the compression cone $\mathcal{C}$ has vertex

$$\boldsymbol{T}_c = \sigma_{cx}\mathbf{e}_1 \otimes \mathbf{e}_1 + \sigma_{cy}\mathbf{e}_2 \otimes \mathbf{e}_2 + \tau_c(\mathbf{e}_1 \otimes \mathbf{e}_2 + \mathbf{e}_2 \otimes \mathbf{e}_1), \tag{20}$$

whose eigenvalues are the value of compressive strength $(\sigma_{c_1}, \sigma_{c_2})$ referred to its principal directions $(\boldsymbol{e}_{c_1}, \boldsymbol{e}_{c_2})$. In its characteristic space, $\mathcal{C}$ can be represented as a plane region as given by Fig. 1(c).

If the vertices of the cones are spherical tensors and $\mathbb{C}$ is isotropic, $\mathcal{K}$ is a spherical set (i.e. $\boldsymbol{T} \in \mathcal{K} \Leftrightarrow \boldsymbol{Q}\boldsymbol{T}\boldsymbol{Q}^T \in \mathcal{K}$ for each rotation $\boldsymbol{Q}$). Then, the characteristic spaces of $\boldsymbol{T} - \boldsymbol{T}_t$, $\boldsymbol{T} - \boldsymbol{T}_s$ and $\boldsymbol{T} + \boldsymbol{T}_c$ coincide so that $\mathcal{K}$ (the intersection among the three cones) and $\boldsymbol{E}^a$ can be represented, in the common characteristic space as a plane domain (Fig. 2(a)).

If the vertices of the cones are not spherical tensors, a plane representation of $\mathcal{K}$ does not exist; however, the 2D *pseudo-representation* depicted in Fig. 2(b) is useful, as it allows an easier determination of the normals to the edges of $\partial\mathcal{K}$. The use of different colours underlines that such representation refers simultaneously to different reference systems.

When the constraints act simultaneously, $\mathcal{K}$ depends on $\boldsymbol{T}_t, \boldsymbol{T}_s$ and $\boldsymbol{T}_c$, and three different situations can occur.

Case 1: The traction constraint is ineffective (Fig. 3), as $\boldsymbol{T}_t$ is outside the shear cone i.e.

$$2\|\boldsymbol{T}_t - \boldsymbol{T}_s\|^2 - (1 + sin^2\phi)tr(\boldsymbol{T}_t - \boldsymbol{T}_s)^2 > 0, \quad tr(\boldsymbol{T}_t - \boldsymbol{T}_s) > 0. \tag{21}$$

Case 2: All the constraints are effective (Fig. 4), when $\boldsymbol{T}_t$ belong to shear cone

$$2\|\boldsymbol{T}_t - \boldsymbol{T}_s\|^2 - (1 + sin^2\phi)tr(\boldsymbol{T}_t - \boldsymbol{T}_s)^2 \le 0, \quad tr(\boldsymbol{T}_t - \boldsymbol{T}_s) \le 0. \tag{22}$$

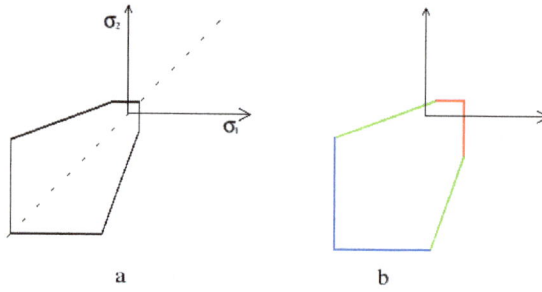

Figure 2: The admissible domain.

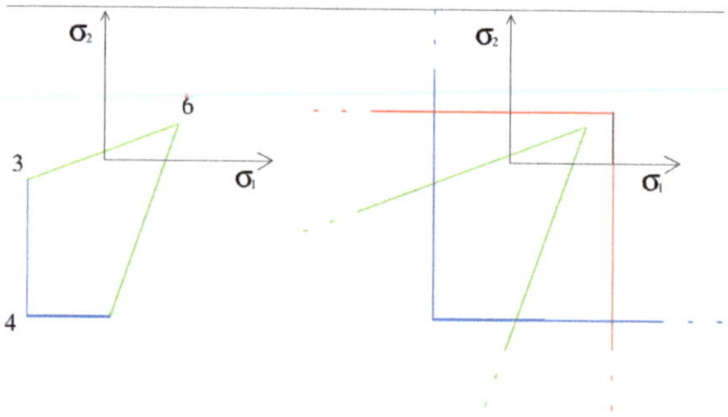

Figure 3: Case 1.

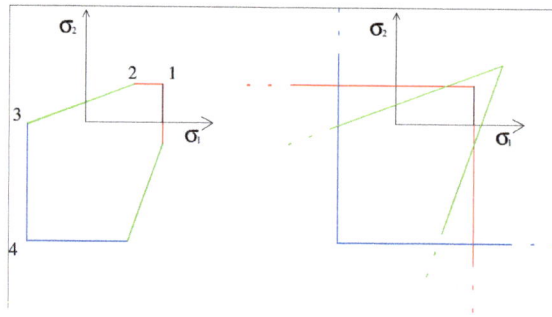

Figure 4: Case 2.

Case 3: the shear constraint is ineffective (Fig. 5), as the intersection between the cones $\mathcal{T}$ and $\mathcal{C}$ is entirely contained in the cone $\mathcal{S}$.

For the sake of example, the semi-explicit procedure to determine the solution of the constitutive equation is given with reference to Case 2.

Let us denote $\mathcal{R}_0$ as the interior of $\mathcal{K}$ and $\mathcal{R}_1$, $\mathcal{R}_{12}$, $\mathcal{R}_2$, $\mathcal{R}_{23}$, $\mathcal{R}_3$, $\mathcal{R}_{34}$ and $\mathcal{R}_4$ the regions that define a partition of $\partial\mathcal{K}$:

$$
\begin{aligned}
\mathcal{R}_0 = \{ \boldsymbol{T} \in \mathrm{Sym} : \det(\boldsymbol{T} - \boldsymbol{T}_t) \geq 0, \, \mathrm{tr}(\boldsymbol{T} - \boldsymbol{T}_t) \leq 0, \\
2\|\boldsymbol{T} - \boldsymbol{T}_s\|^2 - (1 + sin^2\phi)tr(\boldsymbol{T} - \boldsymbol{T}_s)^2 \leq 0, \, tr(\boldsymbol{T} - \boldsymbol{T}_s) \leq 0 \\
\det(\boldsymbol{T} + \boldsymbol{T}_c) \geq 0, \, \mathrm{tr}(\boldsymbol{T} + \boldsymbol{T}_c) \geq 0 \},
\end{aligned}
\tag{23}
$$

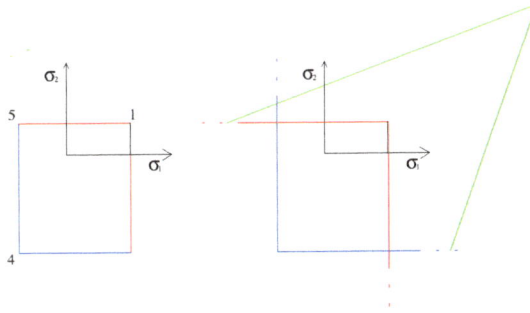

Figure 5: Case 3.

$$\mathcal{R}_1 = \{\boldsymbol{T}_t\}, \tag{24}$$

$$\mathcal{R}_{12} = \{\boldsymbol{T} \in \text{Sym} : \det(\boldsymbol{T} - \boldsymbol{T}_t) = 0, \text{tr}(\boldsymbol{T} - \boldsymbol{T}_t) \le 0,$$
$$2\|\boldsymbol{T} - \boldsymbol{T}_s\|^2 - (1 + sin^2\phi)tr(\boldsymbol{T} - \boldsymbol{T}_s)^2 \le 0, tr(\boldsymbol{T} - \boldsymbol{T}_s) \le 0$$
$$\det(\boldsymbol{T} + \boldsymbol{T}_c) \ge 0, \text{tr}(\boldsymbol{T} + \boldsymbol{T}_c) \ge 0\}, \tag{25}$$

$$\mathcal{R}_2 = \{\boldsymbol{T} \in \text{Sym} : \det(\boldsymbol{T} - \boldsymbol{T}_t) = 0, \text{tr}(\boldsymbol{T} - \boldsymbol{T}_t) \le 0,$$
$$2\|\boldsymbol{T} - \boldsymbol{T}_s\|^2 - (1 + sin^2\phi)tr(\boldsymbol{T} - \boldsymbol{T}_s)^2 = 0, tr(\boldsymbol{T} - \boldsymbol{T}_s) \le 0$$
$$\det(\boldsymbol{T} + \boldsymbol{T}_c) \ge 0, \text{tr}(\boldsymbol{T} + \boldsymbol{T}_c) \ge 0\}, \tag{26}$$

$$\mathcal{R}_{23} = \{\boldsymbol{T} \in \text{Sym} : \det(\boldsymbol{T} - \boldsymbol{T}_t) \ge 0, \text{tr}(\boldsymbol{T} - \boldsymbol{T}_t) \le 0,$$
$$2\|\boldsymbol{T} - \boldsymbol{T}_s\|^2 - (1 + sin^2\phi)tr(\boldsymbol{T} - \boldsymbol{T}_s)^2 = 0, tr(\boldsymbol{T} - \boldsymbol{T}_s) \le 0$$
$$\det(\boldsymbol{T} + \boldsymbol{T}_c) \ge 0, \text{tr}(\boldsymbol{T} + \boldsymbol{T}_c) \ge 0\}, \tag{27}$$

$$\mathcal{R}_3 = \{\boldsymbol{T} \in \text{Sym} : \det(\boldsymbol{T} - \boldsymbol{T}_t) \ge 0, \text{tr}(\boldsymbol{T} - \boldsymbol{T}_t) \le 0,$$
$$2\|\boldsymbol{T} - \boldsymbol{T}_s\|^2 - (1 + sin^2\phi)tr(\boldsymbol{T} - \boldsymbol{T}_s)^2 = 0, tr(\boldsymbol{T} - \boldsymbol{T}_s) \le 0$$
$$\det(\boldsymbol{T} + \boldsymbol{T}_c) = 0, \text{tr}(\boldsymbol{T} + \boldsymbol{T}_c) \ge 0\}, \tag{28}$$

$$\mathcal{R}_{34} = \{\boldsymbol{T} \in \text{Sym} : \det(\boldsymbol{T} - \boldsymbol{T}_t) \ge 0, \text{tr}(\boldsymbol{T} - \boldsymbol{T}_t) \le 0,$$
$$2\|\boldsymbol{T} - \boldsymbol{T}_s\|^2 - (1 + sin^2\phi)tr(\boldsymbol{T} - \boldsymbol{T}_s)^2 \le 0, tr(\boldsymbol{T} - \boldsymbol{T}_s) \le 0$$
$$\det(\boldsymbol{T} + \boldsymbol{T}_c) = 0, \text{tr}(\boldsymbol{T} + \boldsymbol{T}_c) \ge 0\}, \tag{29}$$

$$\mathcal{R}_4 = \{\boldsymbol{T}_c\}. \tag{30}$$

Let $\mathcal{E}_{ij}$ denote the region that will be mapped in $\mathcal{R}_{ij}$ and $\mathcal{E}_0$ the region that will be mapped in $\mathcal{R}_0$.

If $\mathbb{C}\boldsymbol{E} \in \mathcal{K}$ then $\boldsymbol{E} \in \mathcal{E}_0, \boldsymbol{T} = \mathbb{C}\boldsymbol{E}$ and $\boldsymbol{E}^a = \boldsymbol{0}$. Otherwise, $\mathbb{C}\boldsymbol{E}$ has to be projected onto $\partial\mathcal{K}$, and $\boldsymbol{E}$ has to be splitted in two parts $(\boldsymbol{E}^e + \boldsymbol{E}^a)$ so that $\boldsymbol{T}(\in \partial\mathcal{K}) = \mathbb{C}[\boldsymbol{E} - \boldsymbol{E}^a]$.

Firstly, it can be directly verified if $\mathbf{E}$ belongs to one of the regions $\mathcal{E}_0$, $\mathcal{E}_1$ or $\mathcal{E}_4$

$$\mathcal{E}_0 = \{\boldsymbol{E} \in \text{Sym} : \det(\mathbb{C}\boldsymbol{E} - \boldsymbol{T}_t \ge 0, \text{tr}(\mathbb{C}\boldsymbol{E} - \boldsymbol{T}_t \le 0,$$
$$2\|\mathbb{C}\boldsymbol{E} - \boldsymbol{T}_s\|^2 - (1 + sin^2\phi)\text{tr}(\mathbb{C}\boldsymbol{E} - \boldsymbol{T}_s)^2 \le 0, \text{tr}(\mathbb{C}\boldsymbol{E} - \boldsymbol{T}_s) \le 0$$
$$det(\mathbb{C}\boldsymbol{E} + \boldsymbol{T}_c) \ge 0, \text{tr}(\mathbb{C}\boldsymbol{E} + \boldsymbol{T}_c \ge 0\}, \tag{31}$$

$$\mathcal{E}_1 = \{\boldsymbol{E} \in \mathrm{Sym} : \det(\boldsymbol{E} - \mathbb{C}^{-1}\boldsymbol{T_t}) > 0, \mathrm{tr}(\boldsymbol{E} - \mathbb{C}^{-1}\boldsymbol{T_t}) > 0\}$$

or

$$\mathcal{E}_4 = \{\boldsymbol{E} \in \mathrm{Sym} : \det(\boldsymbol{E} + \mathbb{C}^{-1}\boldsymbol{T_c}) > 0, \mathrm{tr}(\boldsymbol{E} + \mathbb{C}^{-1}\boldsymbol{T_c}) < 0\}.$$

If not, let us suppose that $\boldsymbol{E}$ belongs to $\mathcal{E}_{12}$. Let $\boldsymbol{f}_1$, $\boldsymbol{f}_2$ be the orthonormal basis of the characteristic space of $\boldsymbol{T} - \boldsymbol{T}_t$ and let $\theta \in [-\pi/2, \pi/2]$ be the unknown angle that $\boldsymbol{f}_1$ forms with $\boldsymbol{e}_1$ (Fig. 6), so that

$$\boldsymbol{T} - \boldsymbol{T}_t = \sigma \boldsymbol{f}_1 \otimes \boldsymbol{f}_1 \quad \boldsymbol{N}^{12} = \boldsymbol{f}_2 \otimes \boldsymbol{f}_2, \tag{32}$$

where $\sigma \leq 0$ is unknown.

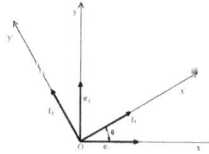

Figure 6: Reference systems.

Since

$$\boldsymbol{f}_1 \cdot \boldsymbol{e}_1 = \boldsymbol{f}_2 \cdot \boldsymbol{e}_2 = cos\theta, \quad \boldsymbol{f}_1 \cdot \boldsymbol{e}_2 = -\boldsymbol{f}_2 \cdot \boldsymbol{e}_1 = sin\theta,$$

we obtain

$$\boldsymbol{f}_1 \otimes \boldsymbol{f}_1 = \frac{1}{1+t^2}\Big(\boldsymbol{e}_1 \otimes \boldsymbol{e}_1 + t(\boldsymbol{e}_1 \otimes \boldsymbol{e}_2 + \boldsymbol{e}_2 \otimes \boldsymbol{e}_1) + t^2 \boldsymbol{e}_2 \otimes \boldsymbol{e}_2\Big), \tag{33}$$

$$\boldsymbol{f}_2 \otimes \boldsymbol{f}_2 = \frac{1}{1+t^2}\Big(t^2 \boldsymbol{e}_1 \otimes \boldsymbol{e}_1 - t(\boldsymbol{e}_1 \otimes \boldsymbol{e}_2 + \boldsymbol{e}_2 \otimes \boldsymbol{e}_1) + \boldsymbol{e}_2 \otimes \boldsymbol{e}_2\Big), \tag{34}$$

where $t = tan\theta$.

The expressions of $\boldsymbol{T}$ and $\boldsymbol{N}^{12}$, with respect to the base $\boldsymbol{e}_1$, $\boldsymbol{e}_2$ can now be deduced by eqns (32)–(34), as functions of the unknowns quantity σ and t, i.e.

$$\boldsymbol{T} = \frac{1}{1+t^2}\Big(\sigma \boldsymbol{e}_1 \otimes \boldsymbol{e}_1 + \sigma t(\boldsymbol{e}_1 \otimes \boldsymbol{e}_2 + \boldsymbol{e}_2 \otimes \boldsymbol{e}_1) + \sigma t^2 \boldsymbol{e}_2 \otimes \boldsymbol{e}_2\Big) \tag{35}$$

$$+ \sigma_{tx} \boldsymbol{e}_1 \otimes \boldsymbol{e}_1 + \sigma_{ty} \boldsymbol{e}_2 \otimes \boldsymbol{e}_2 + \tau_t(\boldsymbol{e}_1 \otimes \boldsymbol{e}_2 + \boldsymbol{e}_2 \otimes \boldsymbol{e}_1), \tag{36}$$

$$\boldsymbol{N}_{12} = \frac{1}{1+t^2}\Big(t^2 \, \boldsymbol{e}_1 \otimes \boldsymbol{e}_1 - t(\boldsymbol{e}_1 \otimes \boldsymbol{e}_2 + \boldsymbol{e}_2 \otimes \boldsymbol{e}_1) + \boldsymbol{e}_2 \otimes \boldsymbol{e}_2\Big). \tag{37}$$

Then, from the system $\boldsymbol{T} = \mathbb{C}[\boldsymbol{E} - \alpha \boldsymbol{N}^{12}]$, three equations can be obtained in the unknowns α, σ and t, where t is the solution of the algebraic equation:

$$at^4 + bt^3 + dt + e = 0 \quad \text{with} \tag{38}$$
$$a = E\phi\epsilon_{12} + \tau_t(\beta\nu^2 - 1),$$
$$b = E(\phi\epsilon_{11} + \beta\epsilon_{22}(\beta\nu^2 + \nu\phi - 1)) + (\beta\nu^2 - 1)(\sigma_{tx}(\beta\nu + \phi) - \sigma_{ty}),$$
$$d = -E\beta(\epsilon_{11}(\beta\nu^2 + \nu\phi - 1) + \epsilon_{22}\phi) - (1 - \beta\nu^2)(\beta\sigma_{tx} - \sigma_{ty}(\beta\nu + \phi)),$$
$$e = -\beta(E\phi\epsilon_{11} + \tau_t(\beta\nu^2 - 1)).$$

Once eqn (38) has been numerically solved,

$$\alpha = \frac{(1+t^2)(E(\epsilon_{11}(t^2 - \beta\nu) + \beta\epsilon_{22}(\nu t^2 - 1)) + (t^2\sigma_{tx} - \sigma_{ty})(\beta\nu^2 - 1))}{E(t^4 - \beta)} \tag{39}$$

and

$$\sigma = \frac{(1+t^2)(E\beta(\epsilon_{11} - t^2\epsilon_{22}) - \beta\sigma_{tx}(1 + \nu t^2) + \sigma_{ty}(t^2 + \beta\nu))}{\beta - t^4} \tag{40}$$

can be determined. Now, if $\alpha \geq 0$ and σ is such that $\boldsymbol{T}$, given by eqn (36), belongs to the intersection of $\mathcal{S}$ and $\mathcal{C}$ (see eqns (10), (11)), then $\boldsymbol{E} \in \mathcal{E}_{12}$ and $\boldsymbol{T}$ is the solution of the constitutive equation. If it does not happen, let us suppose $\boldsymbol{E}$ to belong to $\mathcal{E}_{23}$.

Let $\boldsymbol{g}_1, \boldsymbol{g}_2$ be the orthonormal basis of the characteristic space of $\boldsymbol{T} - \boldsymbol{T}_s$. Then,

$$\boldsymbol{T} - \boldsymbol{T}_s = \sigma\boldsymbol{g}_1 \otimes \boldsymbol{g}_1 + \frac{\sigma(r - m)}{r + m}\boldsymbol{g}_2 \otimes \boldsymbol{g}_2, \tag{41}$$

$$\boldsymbol{N}_{23} = \frac{\sqrt{2}(r - m)}{2s}\boldsymbol{g}_1 \otimes \boldsymbol{g}_1 - \frac{\sqrt{2}(m + r)}{2s}\boldsymbol{g}_2 \otimes \boldsymbol{g}_2, \tag{42}$$

with $r = \sqrt{1 + m^2}$ and $s = \sqrt{1 + 2m^2}$. With respect to the orthonormal basis $\boldsymbol{e}_1, \boldsymbol{e}_2$, from eqns (41), (33) and (34), we obtain

$$\begin{aligned}
\boldsymbol{T} = \frac{-\sigma}{1+t^2}&\left[\left(t^2 - 1 - \frac{2rt^2}{m+r}\right)\boldsymbol{e}_1 \otimes \boldsymbol{e}_1 - \frac{2mt}{r+m}(\boldsymbol{e}_1 \otimes \boldsymbol{e}_2 + \boldsymbol{e}_2 \otimes \boldsymbol{e}_1) \right.\\
&\left. + \left(1 - t^2 + \frac{2r}{m+r}\right)\boldsymbol{e}_2 \otimes \boldsymbol{e}_2\right] + \frac{\tau_{0x}}{m}\boldsymbol{e}_1 \otimes \boldsymbol{e}_1 + \frac{\tau_{0y}}{m}\boldsymbol{e}_2 \otimes \boldsymbol{e}_2 \\
&+ \frac{\tau_s}{m}(\boldsymbol{e}_1 \otimes \boldsymbol{e}_2 + \boldsymbol{e}_2 \otimes \boldsymbol{e}_1),
\end{aligned} \tag{43}$$

$$\begin{aligned}
\boldsymbol{N}_{23} = \frac{\sqrt{2}}{2s(1+t^2)}&[r(1 - t^2) - m(1 + t^2)\boldsymbol{e}_1 \otimes \boldsymbol{e}_1 + rt(\boldsymbol{e}_1 \otimes \boldsymbol{e}_2 + \boldsymbol{e}_2 \otimes \boldsymbol{e}_1) \\
&- m(1 + t^2) - r(1 - t^2)\boldsymbol{e}_2 \otimes \boldsymbol{e}_2]
\end{aligned} \tag{44}$$

in the unknows σ and t. Moreover, with the help of eqn (17), the unknows ω, σ and t can be deduced from the three equations $\boldsymbol{T} = \mathbb{C}[\boldsymbol{E} - \omega\boldsymbol{N}^{23}]$ as in the previous region.

Now, if $\omega \geq 0$ and $\boldsymbol{T}$, as given by eqn (43), belongs to the intersection between $\mathcal{T}$ and $\mathcal{C}$ (see eqns (9), (10)), $\boldsymbol{E}$ belongs to $\mathcal{E}_{23}$ and $\boldsymbol{T}$ is the solution.

Otherwise, let us suppose $\boldsymbol{E}$ to belong to $\mathcal{E}_2$. If so, its image $\boldsymbol{T}$ has to belong to the intersection of $\mathcal{R}_{12}$ and $\mathcal{R}_{23}$. Then, $\boldsymbol{T} - \boldsymbol{T}_t$ can be write with respect to the orthonormal base $\boldsymbol{f}_1, \boldsymbol{f}_2$ as

$$\boldsymbol{T} - \boldsymbol{T}_t = \sigma_1\boldsymbol{f}_1 \otimes \boldsymbol{f}_1, \boldsymbol{N}^{12} = \boldsymbol{f}_2 \otimes \boldsymbol{f}_2, \tag{45}$$

and, with respect to the base $\boldsymbol{g}_1, \boldsymbol{g}_2$ as

$$\boldsymbol{T} - \boldsymbol{T}_s = \sigma_2\boldsymbol{g}_1 \otimes \boldsymbol{g}_1 + \frac{\sigma_2(r - m)}{r + m}\boldsymbol{g}_2 \otimes \boldsymbol{g}_2,$$

$$\boldsymbol{N}_{23} = \frac{\sqrt{2}(r - m)}{2s}\boldsymbol{g}_1 \otimes \boldsymbol{g}_1 - \frac{\sqrt{2}(m + r)}{2s}\boldsymbol{g}_2 \otimes \boldsymbol{g}_2. \tag{46}$$

Let $\theta_1, \theta_2 \in [-\pi/2, \pi/2]$ be the angles between e_1 and, respectively, f_1, g_1. Then, from eqn (45)

$$T = \frac{1}{1+t_1^2} \left(\sigma_1 e_1 \otimes e_1 + \sigma_1 t_1 (e_1 \otimes e_2 + e_2 \otimes e_1) + \sigma_1 t_1^2 e_2 \otimes e_2 \right) \tag{47}$$

$$+ \sigma_{tx} e_1 \otimes e_1 + \sigma_{ty} e_2 \otimes e_2 + \tau_t (e_1 \otimes e_2 + e_2 \otimes e_1), \tag{48}$$

$$N_{12} = \frac{1}{1+t_1^2} \left(t_1^2 e_1 \otimes e_1 - t_1 (e_1 \otimes e_2 + e_2 \otimes e_1) + e_2 \otimes e_2 \right) \tag{49}$$

and from eqn (46)

$$T = \frac{-\sigma_2}{1+t_2^2} \left[\left(\frac{2rt_2^2}{-m-r} + t_2^2 - 1 \right) e_1 \otimes e_1 + \frac{-2mt_2}{r+m} (e_1 \otimes e_2 + e_2 \otimes e_1) \right.$$
$$\left. + \left(1 - t_2^2 - \frac{2r}{-m-r} \right) e_2 \otimes e_2 \right] + \frac{\tau_{0x}}{m} e_1 \otimes e_1 + \frac{\tau_{0y}}{m} e_2 \otimes e_2$$
$$+ \frac{\tau_s}{m} (e_1 \otimes e_2 + e_2 \otimes e_1), \tag{50}$$

$$N_{23} = \frac{\sqrt{2}}{2s(1+t_2^2)} [-m(1+t_2^2) + r(1-t_2^2) e_1 \otimes e_1 + rt_2 (e_1 \otimes e_2 + e_2 \otimes e_1)$$
$$- m(1+t_2^2) - r(1-t_2^2) e_2 \otimes e_2]. \tag{51}$$

with $t_1 = \tan \theta_1$ and $t_2 = \tan \theta_2$. Therefore, by a comparison between eqns (47) and (50), we have

$$\sigma_1 = \frac{1+t_1^2}{m} \frac{m^2 c_1 + m(rc_1 - c_2) + rc_2}{m(1-t_1^2)(1+t_2^2) - r(1+t_1^2)(1-t_2^2)}, \tag{52}$$

with $c_1 = \sigma_{tx} t_2^2 - \sigma_{ty}$ and $c_2 = \tau_{0y} - \tau_{0x} t_2^2$,

$$\sigma_2 = \frac{1+t_2^2}{m} \frac{m^2(t_1^2 - 1)(trT_t - trT_S) + r(1+t_1^2)(\tau_{0y} - \tau_{0x} + m(\sigma_{tx} - \sigma_{ty}))}{m(1-t_1^2)(1+t_2^2) - r(1+t_1^2)(1-t_2^2)} \tag{53}$$

and

$$m^2(\sigma_{tx} t_2 (1-t_1)^2 + \sigma_{ty}(t_1^2 t_2 + 2t_1 - t_2)) + m(rt_2(\sigma_{tx} - \sigma_{ty})(1+t_1^2) - t_1^2 t_2 (\tau_{0x} + \tau_{0y})$$
$$+ 2t_1(t_2^2 \tau_{0x} - \tau_{0y} + t_2(\tau_{0x} + \tau_{0y})) + rt_2^2 (1+t_1^2)(\tau_{0y} - \tau_{0x}) = 0. \tag{54}$$

Moreover, as E^a is the linear combination $E^a = \alpha N_{12} + \omega N_{23}$ and $\mathbb{C}[E - E^a] = T$, three equations are deduced that, together with eqn (54), allow us to solve the problem in the unknowns t_1, t_2, α and ω.

If $\alpha \geq 0, \omega \geq 0$ and $T \in \mathcal{C}$, E belongs to $\mathcal{E}_2$ and T is its image in $\partial\mathcal{K}$. Otherwise, with a procedure similar to the previous one, the (unique) projection T of $\mathbb{C}E$ onto $\mathcal{K}$ can be found.

5 NUMERICAL EXAMPLE

A very simple example is presented to highlight the potentials of the model, capable to accounting for the different directions material's strength. It takes into account solely the traction constraint and despite its simplicity, it allows to appreciate the capabilities of the proposed model generalization. The different directions material's stiffness have also been neglected since the focus is pointed out on material's strength. The results obtained via the

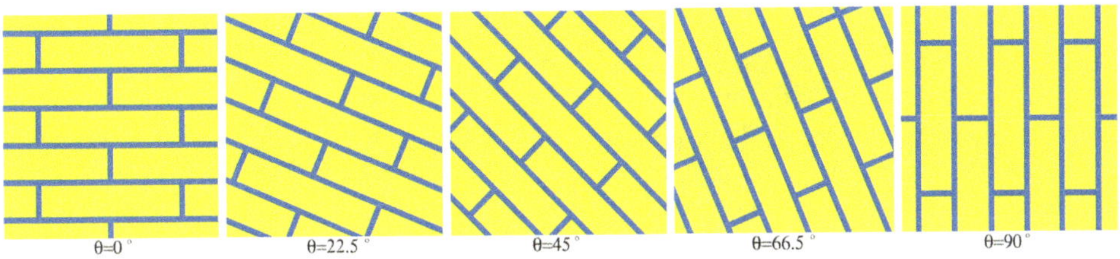

Figure 7: Samples considered in the numerical tensile tests.

proposed model is compared to those obtained by modelling at the micro-scale the texture of a brick's wall.

In the MADY code, the micro-scale models have been built by realizing suitable meshes and by assigning an isotropic non-linear elastic model to the mortar (blue) while the bricks (yellow) have been considered linear elastic – with $E = 0.1$ GPa, $\nu = 0.1$, $\sigma_t = 0.01$ MPa, $\sigma_c = 0.1$ MPa for the mortar, and $E = 0.7$ GPa, $\nu = 0.1$ for bricks (see Fig. 7). These values of mechanical properties are usually used for mortar and bricks in masonry panels. It should however underline that the choice of suitable parameters values is beyond of the scope of the paper. In fact, the main purpose of the example presented is to highlight the model's capability of simulating the texture's effect in masonry structures. For the numerical simulations, 4-nodes plane stress isoparametric elements have been used.

Five numerical tensile tests in the horizontal direction have been performed under displacement control, and the collapse load of the various samples has been determined.

For the "homogenous" model that no longer distinguishes between mortar and bricks, a Young modulus equal to that of the mortar has been assumed; as regards to the strength, the values of σ_{tx} and σ_{ty} have been calibrated by a comparison with the results of the test for $\theta = 0°$ and $\theta = 90°$. All the other cases have been modelled keeping unchanged the values of the mechanical parameters but assuming the appropriate rotations of $\mathcal{K}$. Thus, the generic value of $\mathbf{T}_t$ can be obtained as a function of σ_{tx}, σ_{ty} and θ, i.e.

$$\mathbf{T}_t = \sigma_{tx}\cos^2\theta + \sigma_{ty}\sin^2\theta\mathbf{e}_1 \otimes \mathbf{e}_1 + \sigma_{tx}\sin^2\theta + \sigma_{ty}\cos^2\theta\mathbf{e}_2 \otimes \mathbf{e}_2$$
$$+ (\sigma_{ty} - \sigma_{tx})\sin\theta\cos\theta(\mathbf{e}_1 \otimes \mathbf{e}_2 + \mathbf{e}_2 \otimes \mathbf{e}_1). \tag{55}$$

The graph of Fig. 8 shows the collapse stress vs θ, as predicted by the proposed model compared to that obtained by means of the micro-scale tests.

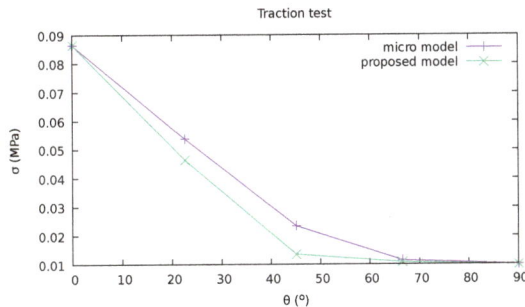

Figure 8: Collapse stress σ vs θ as predicted by the proposed model and the micro-model.

6 CONCLUSION

The proposed model allows to analyze plane bodies made of nonlinear elastic materials whose stress is forced to belong to the intersection of some closed and convex cones. In the proposed formulations, the vertices of the cones are allowed not to be spherical tensors. With this generalization, it has become possible to assign different strength in different directions. In this manner, the model is capable to capture some key aspects of the mechanical behavior of a masonry wall, that generally require to resort to more complex micro scale description. Nevertheless, given the lost of coaxiality between the stress and the inelastic strain becomes more difficult to solve the constitutive equation and some numerical escamotage are needed to have a efficient algorithm. Finally, the capabilities of differentiating the material's strength is a fundamental step of considering different damage process in different directions.

REFERENCES

[1] Del Piero, G., Constitutive equations and compatibility of the external loads for linear elastic masonry-like materials. *Meccanica*, **24**, pp. 150–162, 1989.

[2] Lucchesi, M., Pintucchi, B. & Zani, N., Masonry-like material with bounded shear stress. *Eur. J. Mech. A Solids*, **72**, pp. 329–340, 2018.

[3] Lucchesi, M., Pintucchi, B. & Zani, N., Bounded shear stress in masonry-like bodies. *Meccanica* **53**(7), pp. 1777–1791, 2018.

[4] Lucchesi, M., Padovani, C., Pasquinelli, G. & Zani, N., *Masonry Constructions: Mechanical Models and Numerical Applications*, Springer, 2008.

[5] Berto, L., Saetta, A., Scotta, R. & Vitaliani, R., Shear behaviour of masonry panel: parametric FE analyses. *International Journal of Solids and Structures*, **41**, pp. 4383–4405, 2004.

[6] Pelá, L., Cervera, M. & Roca, P., Continuum damage model for orthotropic materials: Application to masonry. *Computer Methods in Applied Mechanics and Engineering*, **200**(9–12), pp. 917–930, 2011.

[7] Lucchesi, M., Pintucchi, B. & Zani, N., Orthotropic plane bodies with bounded tensile and compressive strength. *Journal of Mechanics of Materials and Structures*, **13**(5), pp. 691–701, 2021.

[8] Lucchesi, M., Pintucchi, B. & Zani, N., Intersection of convex cones as stress range for plane normal elastic bodies, *Proceedings of 1st International Conference on Structural Damage Modelling and Assessment*, 2021.

[9] Šilhavý, M., *Mathematics of The Masonry-like Model and Limit Analysis*, CISM Courses and Lectures, pp. 29–69, 2014.

[10] Mallik, P. K., *Fiber-Reinforced Composites*, M. Dekker, Inc.: New York, 1988.

OPTIMIZING A LIFE CYCLE ASSESSMENT-BASED DESIGN DECISION SUPPORT SYSTEM TOWARDS ECO-CONSCIOUS ARCHITECTURE

MAHMOUD GOMAA, TAREK FARGHALY & ZEYAD EL SAYAD
Faculty of Engineering, Alexandria University, Egypt

ABSTRACT

Within the rapid growth of the energy demand of buildings, cities start to look at ways to shift towards more sustainable solutions that seek the reduction of energy consumption. In the last few decades, Egypt has witnessed a high rate of residential sector investments to accommodate the population inflation. As a result, such buildings in the residential sector consume the highest rates of energy exhaustion to meet the requirements of heating, cooling, and lighting; with the largest amount of burdens, the environment has to afford. Consequently, it is essential to consider energy control and careful analysis of environmental impacts as an essential part of the design of residential buildings. Life Cycle Assessment has gained significant attention in the study of energy control. It helps to analyze the energy patterns and environmental impacts of every single parameter engaged in the design of buildings. However, its complexity limits its integration into the conventional design process. Which lead to the need of engaging computer-aided design techniques and parametric approaches for the easiness of application. This research aims at developing a framework that achieves a reasonable integration between LCA and the traditional design process focusing on early design stages. It provides architects and designers with a structured methodology that enables them to achieve sustainability goals in their designs. The study follows a framework that firstly examines previous research on LCA. Secondly, it highlights the early design decisions and measures their effect on the final output using parametric tools. Lastly, it examines the validation of the developed framework by the implementation of a selected case study. This helps to carry out design optimization based on LCA in the design process.

Keywords: sustainability, LCA, environmental impacts, early design stages, parametric approach.

1 INTRODUCTION

In the last few decades, Egypt has witnessed a rapid rate of increase in the investments of residential buildings due to the population growth and the concentration of the majority of inhabitants in the Nile Delta of Egypt. The Egyptian population is now estimated as 102,976,017 by the Central Agency for Public Mobilization and Statistics (CAPMAS) as of Monday, November 2, 2020 [1].

To face this challenge many expansions have been made in the residential sector to accommodate this inflation, with the focus on building quantities, not the quality of living. Most of these buildings were erected without paying attention to the environmental consideration at the early stages, which led to an excess of energy consumption due to active air-conditioning to afford thermal comfort and well-lit indoor spaces.

As a result, building energy consumption has been increasing to meet the requirements of cooling and electric lighting. Fig. 1 shows that 39.6% of total electrical energy consumption goes to the residential sector in Egypt. This ratio is relatively high when compared to industry or other purposes due to the continued urban expansion under the current conditions in the country and the ongoing increase in the use of electrical appliances, especially air conditioners due to high temperatures during summer.

WIT Transactions on Engineering Sciences, Vol 130, © 2021 WIT Press
www.witpress.com, ISSN 1743-3533 (on-line)
doi:10.2495/CMEM210041

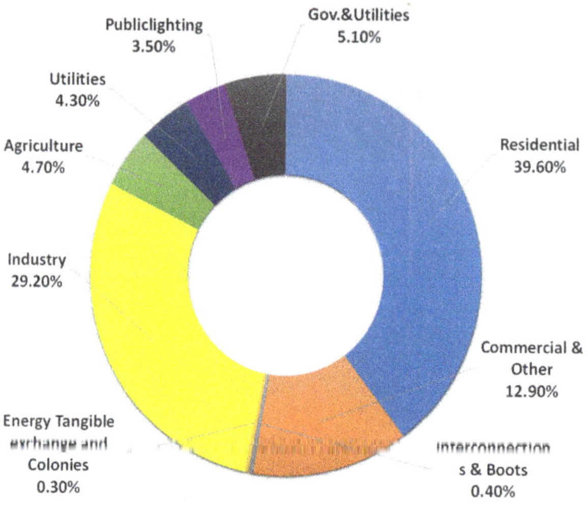

Figure 1: Electricity energy distribution [2].

In addition, the residential buildings in Egypt compromise more than 70% of the building sector which makes them an essential part of achieving positive outcomes of energy-saving plans. The design and configuration of energy-efficient residential units are likely to solve the energy crisis in Egypt [3]. Such critics require interventions based on the environmental point of view that seeks to improve the quality of life while reducing the negative effects on the environment.

Consequently, more sustainable construction solutions have to be integrated into the industry of residential sector. Agenda 21 on sustainable construction defines it as a complete approach that aims to preserve the relationship between the natural and built environment and build settlements that protect human dignity and motivate equity. It is a practice that aims to raise the quality of life for residents by maintaining a balanced relation between the different demands of people and affordable possibilities [4]. Nevertheless, as for developing countries, the application of sustainability guidelines in the construction sector faces multiple barriers such as poverty, technological inertia, low urban investment, inadequate data, lack of interest in the application of sustainability, and the absence of integrated studies [4].

This paper aims at investigating the importance of applying LCA as a tool to assist architects and designers in the evaluation of environmental impacts starting from the early design stages. It offers a methodology of the integration between LCA and the conventional design process. To test the hypothesis of the developed methodology, the research provides an analysis of a case study in Alexandria, Egypt. It provides simulations and analysis of energy for the selected case study and argues that early design decisions would lead to lower consumption of energy and better environmental performance.

2 LITERATURE REVIEW

2.1 Life cycle assessment

Since architects have been increasingly interested in integrating environmental aspects and energy control into their designs, many tools, approaches, and methods have been created for application in the architectural design process. Energy analysis and modeling tools help to predict environmental impacts and calculate needed energy for buildings as well as reducing operational energy. LCA is one of these tools, it allows architects and designers to examine the environmental behavior of buildings procurement, construction, operation, and decommissioning.

As mentioned by Bayer et al. [5], LCA is believed to aid the architectural design decision-making process. LCA is now being developed as a tool to analyze the environmental impact to control and reduce the negative effects of the process. Besides, it is considered essential for a wide range of professions including ecologists, chemists, and all other fields that seek to understand and reduce the negative impacts of any manufacturing process.

Based on the conceptualization of ISO-14040, LCA consists of four phases as shown in Fig. 2; (1) *goal and scope definition*: In this phase, some issues should be cleared, such as the reason for applying LCA study, the application areas of the LCA results, the function of the product system, the product system boundaries, and data category [5], (2) *Life Cycle Inventory Analysis (LCI)*: This phase involves data collection and calculation to quantify inputs and outputs of materials and energy associated with a product system under study, (3) *Life Cycle Impact Assessment (LCIA)*: It consists of several elements classification, characterization, normalization, and weighting, and (4) *Interpretation*: The quantitative results coming from steps 2 and 3 are interpreted qualitatively in order to identify significant issues.

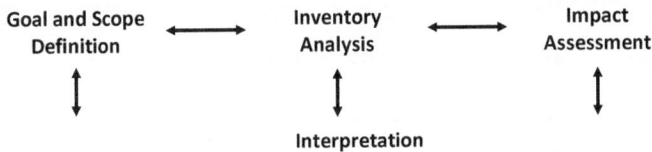

Figure 2: Phases of LCA [6].

Environmental impact categories are measurements of the amount of emissions compared to the environmental impacts that they cause. The impact is always given as a ratio of the quantity of the impact per functional unit of produced product (building or materials). Each category is an indicator of the contribution of a product to a specific environmental problem. These categories are defined by the Life Cycle Impact Assessment (LCIA) methods described below. A set of the environmental impact categories to many LCA methods are provided in Table 1 [5].

LCA can be applied to any product or service, including buildings, to evaluate the environmental impacts of their life cycles [7]. More specifically, the LCA framework has been defined for the building and construction sector as a methodology that provides complete guidelines for applying LCA in that sector [8]. The life cycle of buildings is divided into three main stages with multiple modules as shown in Table 2.

- *The first stage* includes from module A1 to module A5, this stage is divided into two phases of (1) product phase from module A1 to module A3, it is a cradle to gate stage, meaning all processes until the factory's gate, which contains all the used materials in the building, and (2) construction stage from module A4 and module A5, it considers the transportation of building materials from the factory's gate to the construction site, besides all the construction and installation work in site.

Table 1: Output-related parameters [5].

Impact category	Parameter	Abbreviation	Unit
Climate change	Global warming potential	GWP	kg CO_2-equiv.
Ozone depletion	Ozone depletion potential	ODP	kg R11-equiv.
Acidification of soil and water	Acidification potential	AP	kg SO_2-equiv.
Eutrophication	Eutrophication potential	EP	kg PO_4^{3}-equiv.
Formation of photo oxidants	Photochemical ozone creation potential	POCP	kg CH_4-equiv.
Abiotic resource depletion	Abiotic resource depletion potential element	ADPe	kg Sb-equiv.
Abiotic resource depletion	Abiotic resource depletion potential fossil	ADPf	Mj

Table 2. Stages of LCA [9].

Product stage			Process stage		Use stage							End of life stage			
A1	A2	A3	A4	A5	B1	B2	B3	B4	B5	B6	B7	C1	C2	C3	C4
Raw material supply	Transport	Manufacturing	Transport	Construction	Use	Maintenance	Repair	Replacement	Refurbishment	Operational energy use	Operational water use	Deconstruction	Transport	Waste processing	Disposal

- *The second stage* is called the use phase, it includes modules from B1 to module B7. It covers all the operations linked with the building throughout its lifespan. The modules are named use, maintenance, repair, replacement, refurbishment, operational energy use, and operational water use.
- *The third stage* is called the end-of-life stage, it includes modules from module C1 to module C4. This stage is associated with the building deconstruction, the demolishing, transportation, sorting, and treating wastes, and the final disposal [9].

Although LCA is a holistic approach, it has some limitations that should be clear during its application. The main challenges are (1) potentials instead of absolute values: LCA results are not able to predict precise or absolute environmental impacts, practitioners have to rely on models that are only valid within a certain context [10], (2) place independence: LCA is not able to quantify impacts and risks on the environment at a specific location [11], (3) time independence: The models lack temporal dimensions, making it impossible to specify the point in time that emission occurs [12], (4) limitations of impact categories: the impact categories cover a wide range of environmental aspects, but do not cover all relevant environmental aspects [13], (5) assumptions: although LCA aims to be science-based, it involves many technical assumptions and value choices. These assumptions can have a great influence on the results [14], and (6) uncertainties: LCA involves numerous uncertainties such as parameter uncertainties, model uncertainties, and normative uncertainties [10].

LCA studies cannot be completed with the same level of detail, especially in the construction sector. Users like architects or building designs would like to include environmental aspects in an early stage of a building design but without that amount of required details. Simplified approaches are very essential especially in buildings because they consist of many different products and materials, meaning a full detailed LCA is very difficult [15]. To distinguish between different simplification approaches for the LCA of buildings, the EebGuide introduces three categories with increasing levels of detail: screening LCA, simplified LCA, and complete LCA as shown in Table 3 [16].

Table 3: Types of LCA [16].

	Screening LCA	Simplified LCA	Complete LCA
Number of indicators employed	At least 1 or two indicators	Reduced indicator set	A comprehensive set listed in ILC Handbook
Type of data	Generic LCA data	Generic or average LCA data	Specific LCA (EPDs)
Mandatory life cycle modules	A1–A3, B6, B7	A1–A3, B4, B6, B7, C3, C4, D	A1–A5, B1–B7, C1–C4, D
Mandatory building parts to be included	Roof, exterior walls, load-bearing structure, floor slabs, windows	Roof, exterior walls, load-bearing structure, floor slabs, windows, floor finish, foundation	Roof, exterior walls, load-bearing structure, floor slabs, windows, floor finish, foundation, wall finishes, doors, building services including heating or cooling, lighting, lifts, water system

2.2 Energy calculations

Many different types of data are needed for calculating the LCA of a building. This data can be broadly classified as being required either for the embodied or the operational impact calculation. This classification is clarified in Table 4.

Table 4: Categories of needed data for energy calculations (researcher).

Embodied impact	Operational impact
Environmental data for building materials and services	Energy data for energy carriers
Data on the reference service lives (RSL) of building materials	Data for energy demand calculation Physical properties, climate data, and user data

2.3 Integration between the architectural design process and LCA

According to the Royal Institute of British Architects (RIBA), the major planning process is divided into three stages, namely, pre-design stage, design stage, execution stage. Each stage is divided into substages. The focus of this paper is on the design process which is divided into three main steps. These steps are concept design, developed design, and technical design [17] as illustrated in Table 5, the sub-stages are provided in detail in the following section.

Table 5: Phases of the design process [17].

Pre-design		Design stage			Execution stage		
Strategic definition	Preparation and brief	Concept design	Developed design	Technical design	Construction	Handover and closeout	In use

- The concept design stage: it is a stage where the initial design idea is created, the function's requirements for each zone are organized, and the design tasks are specified by geometric parameters including volume, orientation, etc.
- The developed design stage: it is the stage that includes decisions on the typology of construction – whether a skeleton structure or bearing walls will be used.
- The technical design stage: it is the stage that studies building materials, the specific connections between different materials, and the HVAC systems in addition to the increasing level of details in the building model.

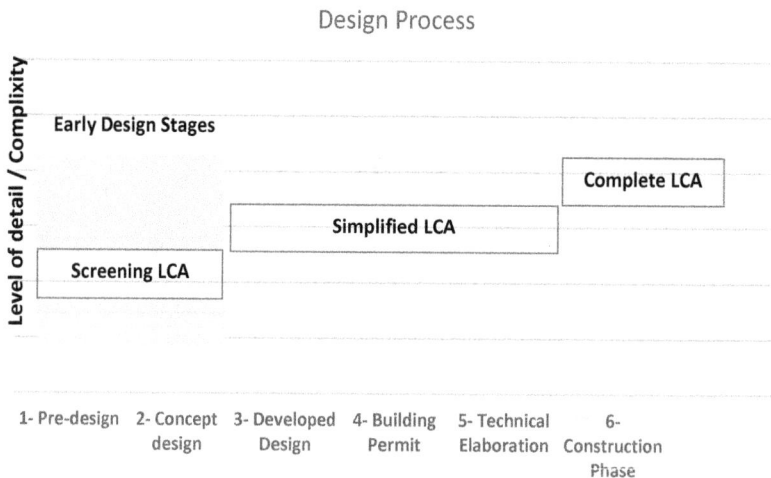

Figure 3: Types of LCA and integration of early design stage [18].

2.4 Methodology

The proposed methodology is created to prove that integrating LCA into the design process is essential to achieve lower environmental impacts. It is divided into five major steps. These steps compromise a workflow that represents an interconnected closed loop as shown in Fig. 4. An explanation of each of these steps is offered as follows:

Step one represents the data input phase which includes all needed parameters to be provided in the early design stage. One of the essential parameters is building geometry. In such a phase of design, it is needed to provide a low detailed geometry model which consists of exterior walls, interior walls, roofs, and slabs. This geometry is presented

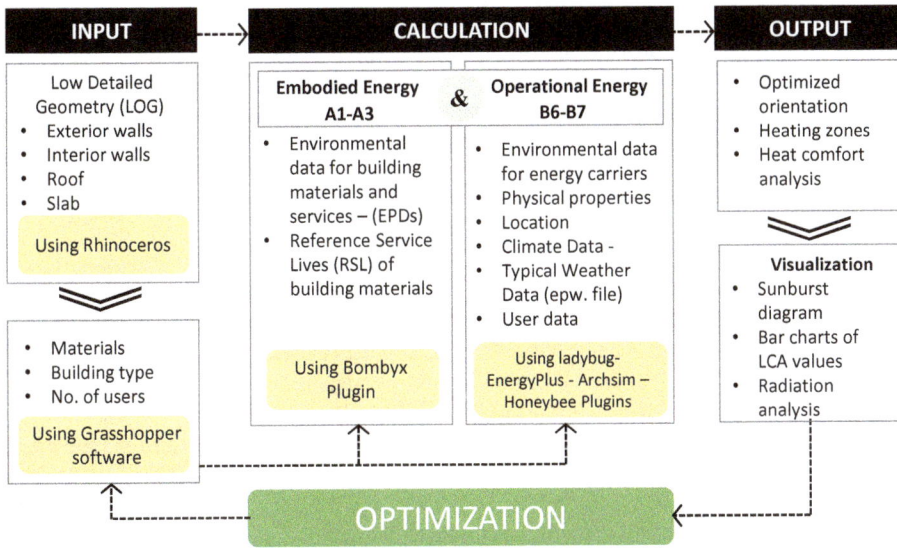

Figure 4: A research framework (researcher).

through a 3D CAD model using a parametric tool such as Rhinoceros 3D. This tool enables designers to alternate and change parameters easily and generate multiple solutions. Other important inputs include assigning materials, building type, and the number of users. Together these inputs provide basic orientation towards the optimization process.

Step two represents the calculation phase. It is divided into two sub-phases of embodied energy calculation and operational energy calculation. On the one hand, embodied energy calculation depends on the environmental data for each material used in the building as the firm which produces this material spreads environmental product declarations (EPDs) files containing this environmental data. It also depends on the reference service life, meaning the lifetime for each material. This data is a part of the database of a Bombyx plugin used to calculate LCA through Grasshopper.

On the other hand, operational energy calculation depends on the environmental data for energy carriers, physical properties of building materials such as density, thermal conductivity, and heat capacity. Besides, it considers building location in order to go with the proper climate data for various building zones. In addition to the user data which represents the need for cooling or drying the indoor comfortability for users. To calculate the operational energy demand, some plugins for Grasshopper are used such as EnergyPlus, Honeybee, Archsim, and ladybug. They are all connected together in order to get results depending on the same database.

Step three represents the outputs, all possible results after the step of energy calculation. These results indicate the energy demand for the building due to the used material, the arrange of comfortability in each zone in the building and the amount of lighting which enters the building, and the amount of heat that faces the building.

Step four represents the visualization phase. This phase is associated with converting all these results into graphs in order to be easy for architects to assess these results. This phase proves easiness in the comparison and optimization process. Most commonly used presentations include bar charts, color-scaled heat zoning charts, or sunburst diagrams.

Step five represents the optimization phase. It is the final stage that compares all the data integrated into the process starting from the input phase. This comparison tends to examine which of the developed scenarios are best in terms of environmental impact. This step serves as an essential part of the systematic design process as it leads to a better understanding of building behavior before conducting detailed design projects.

2.5 Results and discussion

The developed methodology is implemented in a residential area located in Alexandria city, Egypt. The selected project is one of the recent residential expansions that are built to accommodate the population inflation of Alexandria city. It is located on the western edge of the city. The results reflect the steps of the developed methodology. Firstly, the examination of initial geometries using Rhino reveals that the existing orientation of building geometry leads to an inflation of energy consumption due to high amounts of radiation as shown in Table 6. An optimized solution for orientation is to rotate the building by 90° to reduce the amount of radiation and energy consumption accordingly.

Table 6: Optimization of building orientation using a low detailed geometry model.

1	2	The optimized solution
Radiation analysis = 1,291,914 kW Sunlight hours = 3,674,155 hours	Radiation analysis = 1,282,525 kW Sunlight hours = 3,647,510 hours	
3	4	
Radiation analysis = 1,261,082 kW Sunlight hours = 3,618,536 hours	Radiation analysis = 1,215,078 kW Sunlight hours = 3,574,127 hours	Radiation analysis = 1,167,770 kW Sunlight hours = 3,549,795 hours

The existing orientation is directed as trail no 1, in this case, the total amount of solar radiation on the outer surface of the building during the summer zone is 1,291,914 kW. Where the number of hours of direct sunlight received by the building is 3,674,155 hours. These values are decreasing when rotating the building (refer to Table 6). A clear reduction is clear starting from trail no 2 after rotating the building by 30°, then trail no 3 after rotating the building 45°, and reaching a rotation by 60° in trial no 4. The least amount of radiation analysis and in sunlight hours is reached when the building is rotated to be 90°. From these results, building orientation is one of the most important factors which has a significant effect on designing a building and has a noticeable impact on the energy demand of the building to realize the comfortability for building users.

After getting the optimized solution in terms of orientation, it is time to compare between the best solution and the existing one in order to know how important the direction parameter affects the quality of the design and the effectiveness of the outer surroundings. A new simulation is applied to two cases of different orientations. This simulation is called radiation analysis as it allows to calculate the radiation falling on the input geometry using a sky matrix from the select sky mxt component. This type of radiation study is useful for building surfaces, such as windows, where it might be interesting in the energy that can be collected.

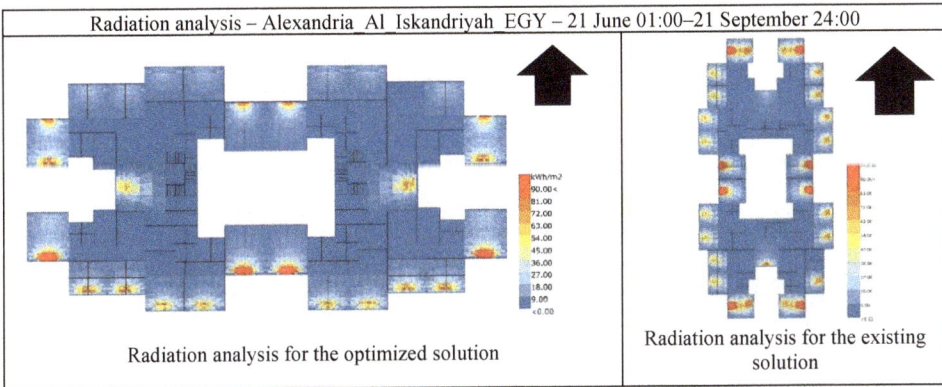

Figure 5:   Radiation analysis of two distinct solutions that vary in terms of orientation (researcher).

It is clear from the above simulation that heat gain in the exciting orientation of the building is higher than the optimized solution. This means that there would be higher energy consumption in the exciting case to exchange this amount of heat with cooling to achieve thermal comfort in indoor spaces. On the other hand, the optimized solution would require a lower energy amount due to lower amounts of solar radiation.

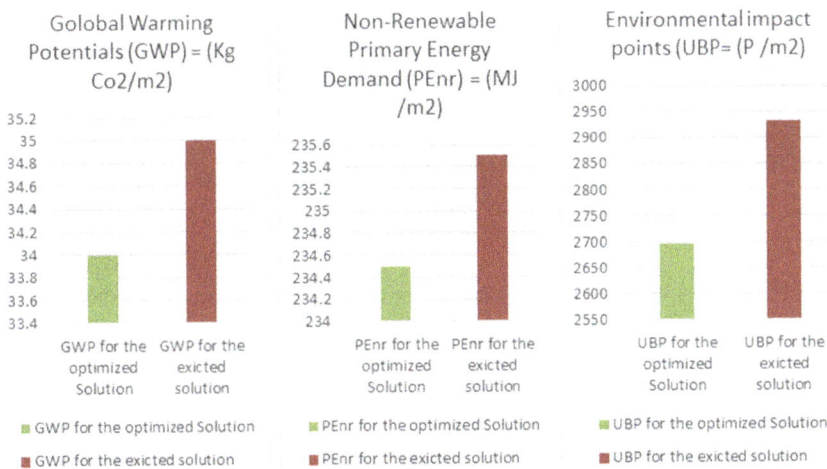

Figure 6: A comparative analysis of LCA values two scenarios (researcher).

In order to adapt the exciting solution, a set of modifications are offered to minimize the heat gain of the building to reduce the consumed energy for cooling. These modifications include adding a double wall thickness in the east and west facades. In addition to a thermal insulation material to stop gaining heat inside spaces. After applying these modifications and getting a satisfying result of energy an LCA calculation has been made to compare between the two cases and to measure the impact of the used materials on the outer surroundings. This reveals that even an adapted solution would lead to higher impacts on the environment. Hence, applying LCA at the early stages of design is essential to reduce environmental impacts.

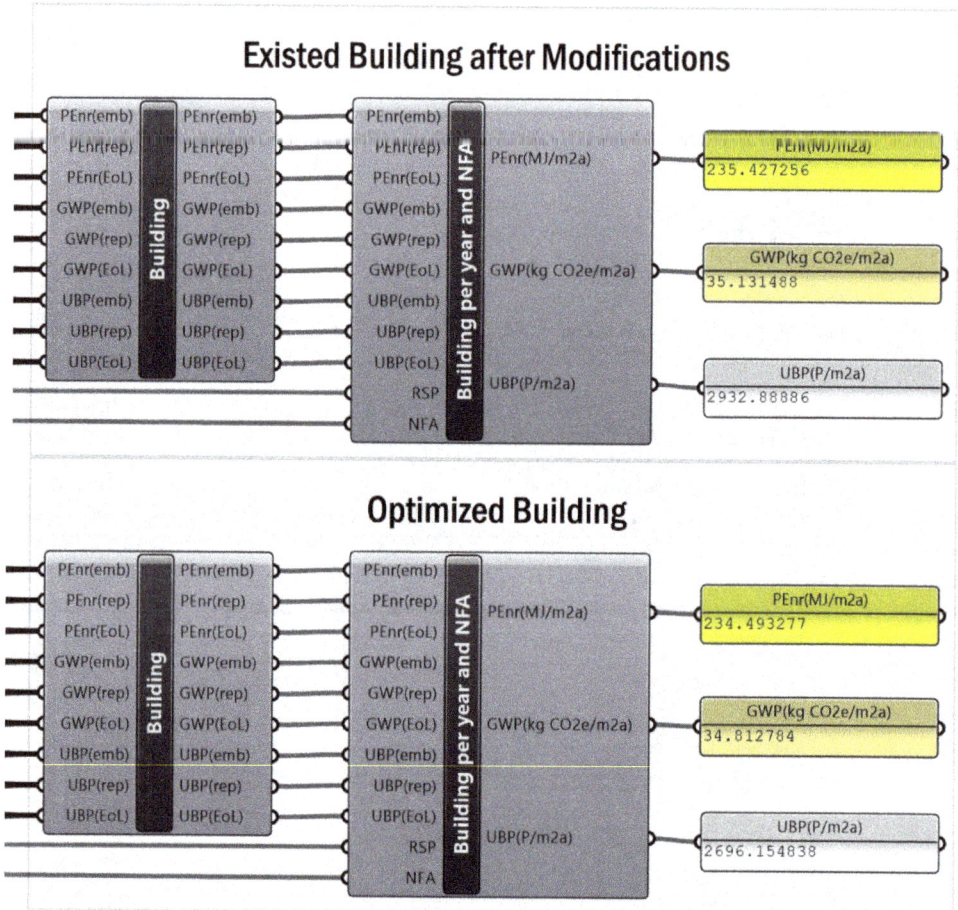

Figure 7: Comparison between the existing building (with proposed modifications) and the optimized solution (researcher).

3 CONCLUSION

LCA has become an increasingly important topic in the design and construction of buildings. It is agreed that LCA is essential to be integrated into the design process. However, its application by designers and architects is not yet widely employed due to complexity and time consumption. This research provided a methodology that aims at integrating LCA in the early stages of the architectural design process. It seeks to achieve a simple framework that architects can easily follow when conducting new designs in order to reduce energy demand and environmental impacts accordingly. This methodology tends to minimize the effort of performing LCA for buildings after finishing it by introducing the interconnection between a screening LCA into the design process in the early design stage. Since design decisions are very important in the early design stages, reducing the environmental impacts became an essential matter to be considered throughout designing a building especially the residential buildings.

REFERENCES

[1]   CAPMAS, Population estimation by the Central Agency for Public Mobilization and Statistics. 2020. https://www.capmas.gov.eg/.
[2]   MERE, The Ministry of Electricity and Renewable Energy of Egypt annual energy report. 2019.
[3]   Aldali, K.M. & Moustafa, W.S., An attempt to achieve efficient energy design for high-income houses in Egypt: Case study: Madenaty city. *International Journal of Sustainable Built Environment*, **5**(2), pp. 334–344, 2016.
[4]   CIB U-I, *Agenda 21 for Sustainable Construction in Developing Countries*, Conseil International du Batiment: Rotterdam, 2002.
[5]   Bayer, C., Gamble, M., Gentry, R. & Joshi, S., *AIA Guide to Building Life Cycle Assessment in Practice*, The American Institute of Architects: Washington, DC, 2010.
[6]   ISO14040. Environmental management – Life cycle assessment – Principles and framework (ISO 14040:2009), 2009.
[7]   Frostell, B., Life cycle thinking for improved resource management: LCA or? 2013.
[8]   Kotaji, S., Schuurmans, A. & Edwards, S., *Life-Cycle Assessment in Building and Construction: A State-of-the-Art Report, 2003*, Setac, 2003.
[9]   Lasvaux, S. et al., Achieving consistency in life cycle assessment practice within the European construction sector: The role of the EeBGuide InfoHub. *The International Journal of Life Cycle Assessment*, **19**(11), pp. 1783–1793, 2014.
[10] Klöpffer, W. & Grahl, B., *Life Cycle Assessment (LCA): A Guide to Best Practice*, John Wiley & Sons, 2014.
[11] König, H. & De Cristofaro, M.L., Benchmarks for life cycle costs and life cycle assessment of residential buildings. *Building Research and Information*, **40**(5), pp. 558–580, 2012.
[12] Collinge, W.O., Landis, A.E., Jones, A.K., Schaefer, L.A. & Bilec, M.M., Dynamic life cycle assessment: Framework and application to an institutional building. *The International Journal of Life Cycle Assessment*, **18**(3), pp. 538–552, 2013.
[13] Finnveden, G., On the limitations of life cycle assessment and environmental systems analysis tools in general. *The International Journal of Life Cycle Assessment*, **5**(4), p. 229, 2000.
[14] Guinée, J.B., Huppes, G. & Heijungs, R., Developing an LCA guide for decision support. *Environmental Management and Health*, 2001.
[15] Wittstock, B. et al., EeBGuide guidance document part B: Buildings. Operational guidance for life cycle assessment studies of the energy-efficient buildings initiative. pp. 1–360, 2012.
[16] Gantner, J. et al., *EeBGuide Guidance Document Part B: Buildings. Operational Guidance for Life Cycle Assessment Studies of the Energy Efficient Building Initiative*, Fraunhofer Verlag, 2015.

[17] Sinclair, D., *Design Management: RIBA Plan of Work 2013 Guide*, Routledge, 2019.
[18] Kiss, B. & Szalay, Z. (eds), The impact of decisions made in various architectural design stages on life cycle assessment results. *Applied Mechanics and Materials*, 2017.

PARTICLE-BASED FLOW VORTICITY ANALYSIS BY USE OF SECOND-GENERATION WAVELETS

ODDNY H. BRUN[1], JOSEPH T. KIDER JR.[1] & R. PAUL WIEGAND[2]
[1]School of Modeling, Simulation, and Training University of Central Florida, USA
[2]Department of Computer Science and Quantitative Methods, Winthrop University, USA

ABSTRACT

Modeling, simulating, and analyzing turbulent flow is a topic of high interest from both a verification and accuracy aspect. This work presents computational methods and experimental measures of turbulent fluid flow modeled with particle-based smoothed particle hydrodynamics (SPH), as well as the use of second-generation wavelets to analyze the nature of vorticity. Modeling and analyzing vorticity by use of first-generation wavelets for regular grid methods are well presented in literature. We are unaware of any work on this topic for particle-based methods. The difference between regular grid-based and particle-based approaches are due to irregularities introduced by the latter. We found that second-generation wavelets proved to be robust, fast, and reliable. Second-generation wavelets are designed to handle irregular grids and do not rely on a dyadic number of observations, which make them suitable candidates for SPH analysis as opposed to first generation wavelets. The resulting significant discrete wavelet transform (DWT) coefficients are found to be representative of the flow sections that may benefit from additional attention in the simulation model. The robustness of the method allows for fast initial screening of the flow to highlight sections that are of interest for more detailed analysis. Here, robustness refers to the two parameters significance level and grid resolution. Our results are demonstrated using a 2D sloshing tank case.
Keywords: smoothed particle hydrodynamics (SPH), vorticity, second-generation wavelets, thresholding.

1 INTRODUCTION

Real-time simulations of flow models depend on fast and efficient methods for analyzing the vast amount of data typically generated during a simulation run. In this work, we show how a method based on second-generation wavelets was used to analyze the data from a smoothed particle hydrodynamics (SPH) simulation of less smooth flow. While the irregular particle spacing in SPH represents welcoming flexibility it has also proven to be a challenge to the method from the very beginning and still continues to be (Monaghan [1]). The degree of irregularity increases for turbulent flow and hereby challenges the smoothing process that is a fundamental step towards achieving accuracy (Rafiee et al. [2]). The work presented here demonstrates how a second-generation wavelet approach is successfully used to:

- efficiently identify the sections of the flow of relevance for improvement and
- handle SPH generated data of irregular structure.

As pointed out by numerous researchers, among others Farge et al. [3], flow vorticity is an important parameter in less smooth flow modeling. We apply second-generation wavelet analysis to the vorticity values of our flow simulation data. As far as we have been able to verify, no such approach has been presented before. Previous work presented on flow analysis using wavelets is based on first-generation wavelets data on regular grid structures. First-generation wavelets require a dyadic number of observations and a regular grid structure. For particle-based methods, including SPH, these conditions are not met. Hence, the motivation for using second-generation wavelets, which do not require either of those two conditions. With the second-generation wavelet method, we were able to quickly and efficiently identify

WIT Transactions on Engineering Sciences, Vol 130, © 2021 WIT Press
www.witpress.com, ISSN 1743-3533 (on-line)
doi:10.2495/CMEM210051

the sections of the flow where the focus of vorticity analysis is expected to be beneficial. In section two of this paper, the methods are presented. The results of our analysis are presented in section three. Data from the SPH simulation of a two-dimensional sloshing tank case was used. Concluding remarks are presented in section four.

2 THE FUNDAMENTALS IN TERMS OF METHODS AND CONDITIONS

In this section, we have included methods and conditions of specific relevance to the work presented in this paper. For the fundamentals of fluid dynamics, including smoothed particle hydrodynamics (SPH) and wavelet theory, the reader is encouraged to seek the relevant sources on those topics. Example of such sources are Monaghan [1], Lucy [4], Mallat [5], Monaghan [6]. The relevance of vorticity analysis in turbulent fluid modeling is thoroughly discussed by Farge [7], Farge et al. [8] as well as several later publications by the same group of researchers. The motivation for use of wavelets is based on properties related to orthogonal and biorthogonal wavelets that provide a decomposition of the flow field into independent components (Farge et al. [3], Urban [9], Urban [10]).

The difference between first- and second-generation wavelets of relevance to vorticity analysis lies in the fact that a first-generation wavelet can be chosen so that its base functions are spanned by a Riesz basis, meaning it is orthogonal or close to orthogonal, and hence provide an independent component split. Second-generation wavelets do not fulfill the conditions of a Riesz basis, which leads to two concerns: stability and lack of independence in the decomposition. The stability concern is remediated by a check and balance to ensure low condition numbers leading to stable basis functions as presented by Vanraes et al. [11], [12]. The lack of independence in our case was expected to be ignorable as the purpose of our analysis was of the nature of feature extraction. Roussel et al. [13] quantified the effect of biorthogonal wavelets to a loss of *three percent* in enstrophy compared to orthogonal when analyzing a 3D homogeneous turbulent flow, something one should take into consideration if the results were to be used in an operational step of the simulations with influence on mass, momentum, or energy conservation. Under such conditions, a further quantification of the lack of independence would be required. The nature of second-generation wavelets lets the basis functions adapt to the irregularity of the data (Jansen and Oonincx [14], Sweldens [15]), which is part of why orthogonality is not achievable.

The nature of vorticity in turbulent flow is assumed to be comprised of a coherent and an incoherent component. The coherent component is characterized by stronger and further reaching correlation as opposed to the incoherent component that takes the nature of Gaussian white noise. The separation of the two components is achieved by applying a threshold to the coefficients produced by a discrete wavelet transform. This mechanism is the same as what has been known for decades for signal or data denoising. The specifics of selecting the thresholding level is among others described by Donoho and Johnstone [16] and Abramovich et al. [17] as examples of hard and soft thresholding, respectively. We used a hard threshold in accordance with earlier work on turbulence analysis. This threshold, t_{cut}, was calculated as:

$$t_{cut} = \sqrt{\frac{2}{3}\sigma_{cD}^2 ln(n)}$$ (1)

where σ_{cD}^2 is the DWT coefficients' estimated variance and n is the number of coefficients. Multiple previous works have used alternative ways to calculate the estimate, one being the variance of all coefficients, another being the variance of the smaller coefficents, referred to respectively as high and low thresholding in the presentation of our results below. The low threshold is calculated with an iterative process and halted when the difference in variance between two consecutive iterations becomes rather small. The reasoning behind this approach is based on the fact that white noise variance is constant. We found that in most cases four iterations were enough to achieve convergence.

We would in general recommend that a thresholding mechanism should be selected depending on the purpose and sensitivity of the analysis. This is based on results from earlier work (Brun [18]) where soft thresholding resulted in more accurate results than hard when used for discontinuity identification purpose.

The open-source software *DualSPHysics* (Domínguez et al. [19]) was used for our simulations and the discrete wavelet transforms were performed by the code for second-generation wavelets analysis provided in *MATLAB* MathWorks [20] with a symmlet wavelet. The technical details of second-generation wavelets are beyond the scope of this paper, however, open access to MATLAB source code for reproduction purposes will be made available at a later time.

3 RESULTS

In this section, we present the major findings and implications of vorticity analysis by second-generation wavelets. The major findings were:

- The nature of vorticity in particle-based flow simulations were identifiable by the use of second-generation wavelets.
- The coherent vorticity structures were separated from the incoherent ones by the significant discrete wavelet transform (DWT) coefficients.
- The method demonstrated robustness in terms of resolution level needed to locate the coherent structures in the flow.
  - A relatively high threshold for significance was sufficient to identify the location of the stronger coherence. A lower threshold based on the white noise variance may be used to capture more details.
  - A coarse resolution structure relative to average particle distance was able to identify the location of the stronger coherence. A finer resolution closer to average particle distance provided a more accurate location of the coherent structures. A resolution equal to initial particle spacing was insufficient for wavelet analysis.
  - Combining high threshold and coarse resolution structure allowed for a fast approach to identify the sections of the flow where the stronger coherence were located.
- The irregular particle structure that particle based method represents was remediated by the second-generation wavelets.
- The method is memory efficient and fast due to the nature and technique used for the so-called lifting scheme combined with second-generation wavelets.

The results were demonstrated on a 2D sloshing tank case simulated by the smoothed particle hydrodynamics (SPH) method with the following case setup:

A rectangular body of water, initially located at the entire lower part of a tank, gets exposed to gravitational force and an additional external force represented by a sine wave causing the tank to rock back and forth over the y-axis (Fig. 1). The setup was with an initial average particle distance of $dp = 0.001$ m which resulted in $82,708$ fluid particles. Results were sampled at 40 kHz. The simulation ran for 8.35 seconds.

3.1 Coherent vs. non-coherent vorticity separated by second-generation wavelet transform

The vorticity analysis was performed on a time frame basis by applying the discrete wavelet transform (DWT) to the particles' vorticity values. The distinction between non-significant

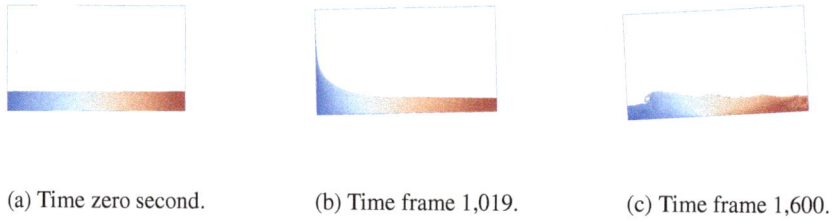

(a) Time zero second. (b) Time frame 1,019. (c) Time frame 1,600.

Figure 1: Sloshing motion case, shown at initial stage, $t = 0$, at $t = 1,019$, $\sim$2.56 sec, and at $t = 1,600$, $\sim$4.02 sec. Colors are based on initial particle position, that is: particles positioned left are blue, to the right are red.

and significant DWT coefficients was obtained by applying a threshold, which meant setting the DWT coefficients with absolute value less than the threshold level to zero while keeping unchanged the DWT coefficients of absolute value higher than the threshold level. The resulting significant DWT coefficients appeared in the sections of the flow where coherent vorticity structures were present. An example from time frame $1,600$, $\sim$4.02 seconds into the simulation run, is presented in Fig. 2. The amount and locations of coherent vorticity seem to be related to the smoothness of the flow, as expected. This was observed when comparing the results of time frame $t = 1,019$, $\sim$2.56 sec., which is an earlier phase of the flow where the forces with which the matter hits the walls are lower and less turbulence present (Fig. 3), compared to the later time frame $1,600$.

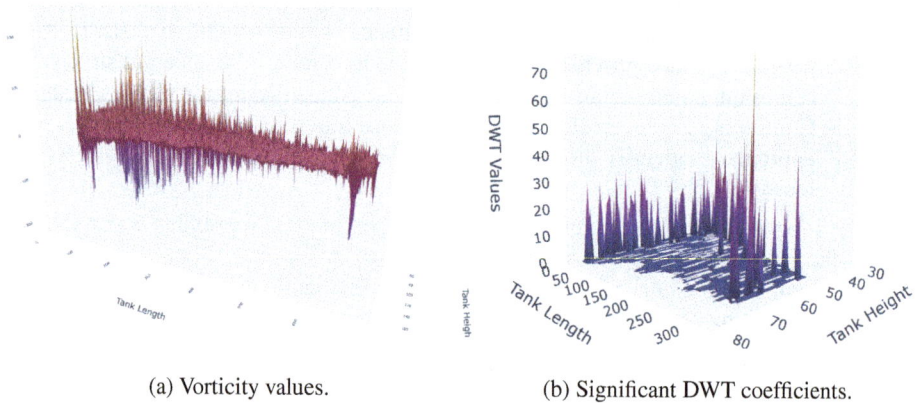

(a) Vorticity values. (b) Significant DWT coefficients.

Figure 2: Vorticity values and significant DWT coefficients indicating coherent vorticity at $t = 1,600$, $\sim$4.02 sec.

3.2 Robustness

The vorticity analysis based on second-generation wavelet analysis demonstrated robustness both in terms of threshold levels as well as resolution used for the flow field. The work presented on regular grid-based vorticity analysis have used the approach of both a higher

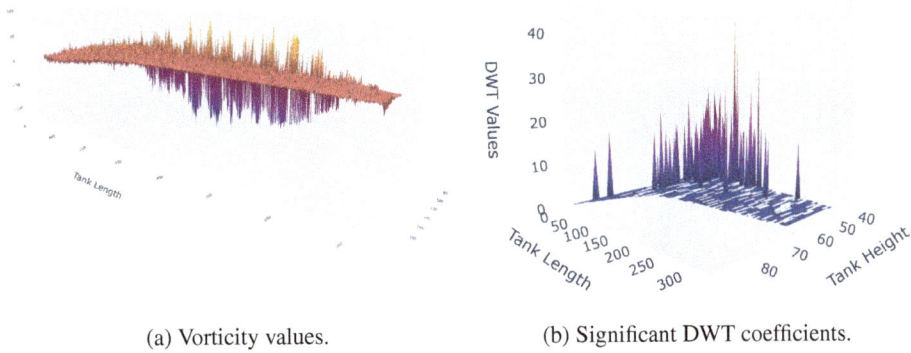

(a) Vorticity values.

(b) Significant DWT coefficients.

Figure 3: Vorticity values and significant DWT coefficients indicating coherent vorticity at $t = 1,019, \sim 2.56$ sec.

threshold based on the total flow field variance as well as a threshold based on the white noise variance estimate which leads to a lower threshold. We found the higher threshold resulted in just above *three percent* significant DWT coefficients while the lower threshold resulted in just above *eleven percent* when comparing results from one and the same time frame.Figs 4 and 5 both visually and quantitatively demonstrate the difference .In general, the selection of threshold should depend on the purpose of the analysis. With a regular grid and first-generation wavelet analysis, it has been reported that the higher threshold represented a somewhat lower level of enstrophy compared to if the lover threshold was used Farge et al. [8], Okamoto et al. [21]. We did observe a similar effect as we performed an inverse discrete wavelet transform (IDWT) of the significant DWT coefficients and compared the results by taking the $l_2 norm$. The higher cut resulted in about *two percent* lower $l_2 norm$ of the reproduced vorticity values than the lower cut. For the purpose of the work here, our choice would be an optimization of accuracy versus computational resources and real-time restriction.

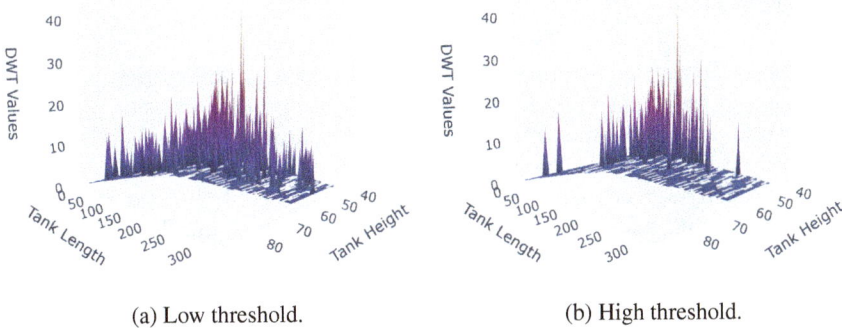

(a) Low threshold.

(b) High threshold.

Figure 4: Threshold robustness. Significant DWT coefficients indicating coherent vorticity values at $t = 1,019, \sim 2.56$ sec of 3.5 and 11.0 percent, respectively.

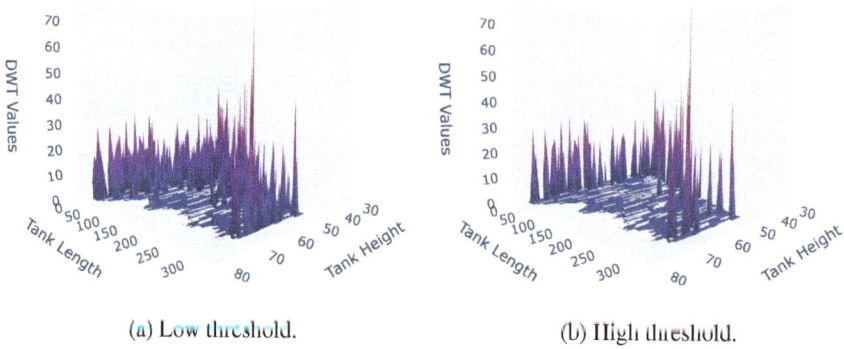

(a) Low threshold. (b) High threshold.

Figure 5: Threshold robustness. Significant DWT coefficients indicating coherent vorticity values at $t = 1,600$, ~ 4.02 sec of 3.1 and 11.0 percent, respectively.

The setup leading to the results presented so far was based on a resolution structure of 0.0013 by 0.0013 m, where the 0.0013 m is the radius of the circle defining the smoothing function's compactly supported area for our 2D SPH based case. A finer resolution equivalent to the initial particle spacing of 0.001 m resulted in too sparse particle representation proving insufficient for the wavelet analysis. However, more coarse structures equivalent to *two* and *three* times the 0.0013 m did identify significant DWT coefficients located in the sections that were already identified by use of the higher resolution as shown in Fig. 6(a) and 6(b), respectively. Further, the reduction of the resolution level by the factors *two* and *three* slightly increased the percentage of significant DWT coefficients to *4.1* and *4.8*, respectively. This increase may partly have been caused by the fact that for lower resolution the vorticity range value for each cell was used as opposed to the actual vorticity value used at the higher resolution. The robustness of this method both in terms of resolution and threshold level demonstrated the potential of using such an approach to identify the sections of the flow that represent strong coherence. By that, the sections that have the higher potential of improved smoothing may be located in a fast and efficient manner.

3.3 Particle irregularity

In SPH, matter is represented in terms of particles where each particle carries its parameters, which typical are mass, position, acceleration, velocity, pressure, density, and vorticity. Additional particle characteristics may be added. The particles are initially in a regular equally-spaced arrangement. As soon as any force is applied to the matter, the particles are no longer equally-spaced. This loss of regularity is observed by a varying number of particles located within a section of given size. For a section of the flow with size equal to the compactly supported domain of the smoothing function used in a particular SPH simulation, the number of particles may vary from zero to several times the number of the initial arrangement. Fig. 7 shows multiple white areas, which means no particles are present in those parts of the flow field. The flow field itself is also of irregular shape with rugged edges. The second-generation wavelet transform was able to handle both those factors, something that would prohibit analysis by a first-generation wavelet.

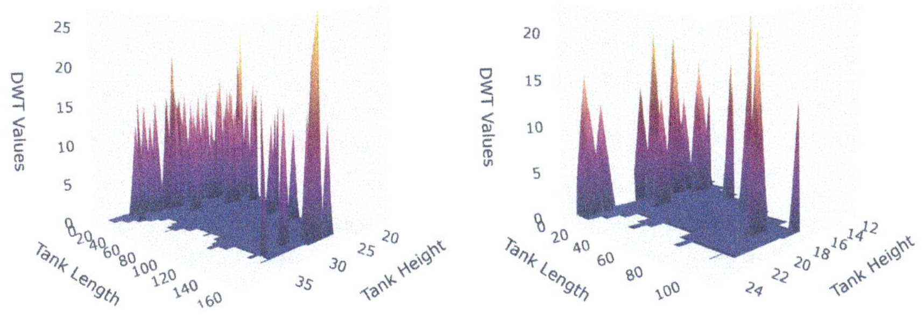

(a) Resolution 0.0026 by 0.0026 m.   (b) Resolution 0.0039 by 0.0039 m.

Figure 6: Resolution robustness. Significant DWT coefficients indicating coherent vorticity at $t = 1,600$, ~4.02 sec of 4.1 and 4.8 percent, respectively, using a high threshold level.

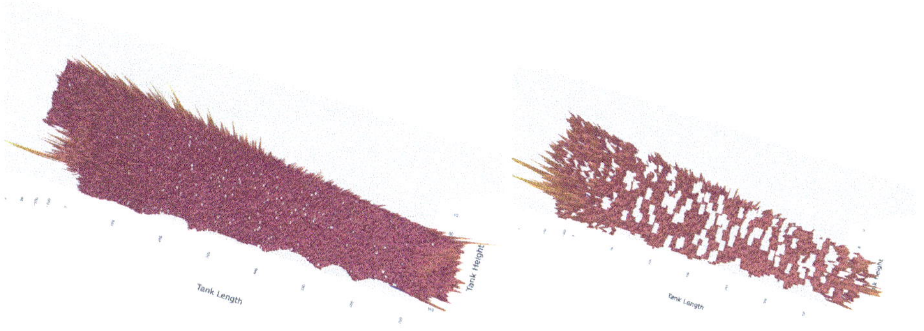

(a) Vorticity values.   (b) DWT coefficients uncut.

Figure 7: Top view of vorticity values and DWT coefficients at $t = 2,896$, ~7.24 sec where white sections within the flow field are sections with no particles present.

3.4 Fast and memory efficient

The amount of data generated by simulating the 2D sloshing tank case according to the setup described in the beginning of this section, a run was in the order of 82,708 fluid particles times $3,340$ time frames times the number of parameters tracked per particle. With such a vast amount of data, efficient feature extraction tools are necessary. Our particular case was motivated by the need to separate the less smooth sequences into those where improvement (such improvements are outside the scope of this paper) was expected to have significant impact in terms of accuracy and those that could be neglected. Each of the *82,708* fluid particles represents such a smoothing sequence. Of those, about *25 to 28 percent* represent less smooth sequences. The vorticity analysis by use of second-generation wavelets did enable us to reduce the number of particles relevant for further attention to about *three percent*, as well as identifying their location in the flow field. Fig. 8(a) and 8(b) demonstrate the feature extraction achieved. Other factors that contributed to speed and memory efficiency were:

- Use of a spatial hashing structure (Hastings et al. [22]);
- Fast wavelet transform algorithms (Mallat [23]);
- Wavelet functions generated by use of lifting scheme (Sweldens [15], [24]); and
- Minimal memory usage as filter are linear and generated in-place (Sweldens [15]).

We used a spatial hashing structure with increasing cell sizes when evaluating the effect of resolution used for the flow field. Fast wavelet transform algorithms combined with low memory usage as the second-generation wavelet are generated by the lifting scheme, a scheme that can speed up the conventional wavelet transform (Sweldens [15], [24]). While none of these other factors were a result of our work, they are listed here as a confirmation of the efficiency of high value when dealing with large data sets.

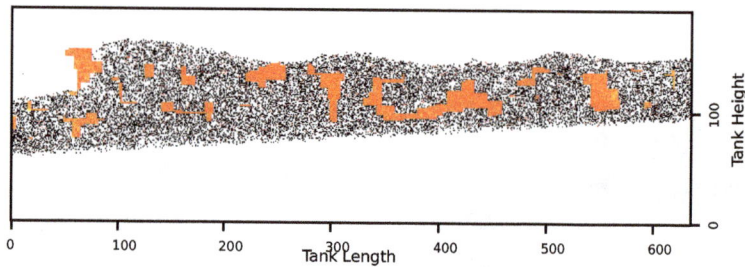

(a) At $t = 1,600$, 3.1 percent sign. coeffs.

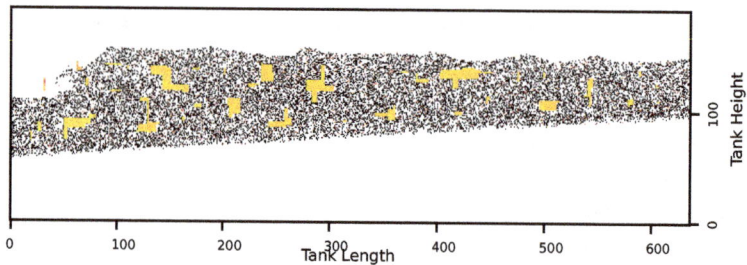

(b) At $t = 2,896$, 3.0 percent sign. coeffs.

Figure 8: Location of coherent vorticities (orange/yellow) compared to discontinuous smoothing sequences (black) for sloshing motion. Tank is divided into cells of 0.0013 by 0.0013 m.

4 CONCLUSION

The result of the work presented here shows the second-generation wavelets' ability to handle irregular data structures and efficiently extract key features of interest from vast amount of non-dyadic data. This was demonstrated by carrying out vorticity analysis of a two-dimensional turbulent flow simulated by smoothed particle hydrodynamics. The differentiation between the significant and non-significant discrete wavelet transform coefficients separated the coherent vorticity components and the incoherent ones. Earlier work states that the coherent components are characterized by stronger and further reaching correlation, so being able to locate those components is of high importance when optimizing were to focus on model improvements under turbulent flow conditions. We found the method robust and reliable as far as we were able to compare to vorticity analysis results by first-generation wavelets on regular grids. The method may have limitations in terms of accuracy as it does not produce orthogonality between the coherent and incoherent components. Such a limitation may be relevant in case the results were to be used as part of a model where the ability to conserve mass, momentum, or energy was influenced. This was not a concern in our work as the feature extraction effect was the goal of our analysis and we achieved a reduction in sections of relevance by a factor of about *nine*.

REFERENCES

[1] Monaghan, J. J., Particle methods for hydrodynamics. *Computer Physics Report*, 1985.
[2] Rafiee, A., Cummins, S., Rudman, M. & Thiagarajan, K., Comparative study on the accuracy and stability of sph schemes in simulating energetic free-surface flows. *European Journal of Mechanics, B/Fluids*, **36**, pp. 1–16, 2012.
[3] Farge, M., Schneider, K. & Kevlahan, N., Non-gaussian and coherent vortex simulation for two-dimensional turbulence using an adaptive orthogonal wavelet basis. *Physics of Fluids*, 1999.
[4] Lucy, L.B., A numerical approach to the testing of the fission theory. *The Astronomical Journal*, **82**, pp. 1013–1024, 1977.
[5] Mallat, S., *A Wavelet Tour of Signal Processing The Sparse Way*, Academic Press, 2009.
[6] Monaghan, J.J., Sph without a tensile instability. *Journal of Computational Physics*, 2000.
[7] Farge, M., Wavelet transforms and their applications to turbulence. *Annual Review of Fluid Mechanics*, 1992.
[8] Farge, M., Schneider, K. & Kevlahan, N.K.R., Coherent structure eduction in wavelet-forced two-dimensional turbulent flow. *IUTAM Symposium on Dynamics of Slender Vortices*, eds E. Krause & K. Gersten, 1998.
[9] Urban, K., Wavelet basis in h(div) and h(curl). *Mathematics of Computing*, 2000.
[10] Urban, K., *Wavelets in Numerical Simulation*, Springer, 2002.
[11] Vanraes, E., Jansen, M., & Bultheel, A., Stabilised wavelet transforms for non-equispaced data smoothing. *preprint Signal Processing stabwt*, 2000.
[12] Vanraes, E., Jansen, M., & Bultheel, A., Stabilised wavelet transforms for non-equispaced data smoothing. *Signal Processing*, 2002.
[13] Roussel, O., Schneider, K., & Farge, M., Coherent vortex extraction in 3d homogeneous turbulence: Comparison between orthogonal and biorthogonal wavelet decomposition. *Journal of Turbulence*, 2005.
[14] Jansen, M. & Oonincx, P., *Second Generation Wavelets and Applications*, Springer, 2010.
[15] Sweldens, W., The lifting scheme: A construction of second generation wavelets. *SIAM Journal on Mathematical Analysis*, 1998.

[16] Donoho, D.L. & Johnstone, I.M., Adapting to unknown smoothness via wavelet shrinkage. *Journal of the American Statistical Association*, **90**, pp. 1200–1224, 1995.

[17] Abramovich, F., Benjamini, J., Donoho D.L. & Johnstone, I. M. Adapting to unknown sparcity by controlling the false discovery rate. *The Annals of Statistics*, **34**, pp. 584–653, 2006.

[18] Brun, O.H. Improved interpolation in sph in cases of less smooth flow. Master's thesis, Institute for Simulation and Training, University of Central Florida, 2016.

[19] Domínguez, J.M., Crespo, A.J.C. & Rogers B.D., Users guide for dualsphysics code, 2016. https://dual.sphysics.org/.

[20] MathWorks, Lifting method for constructing wavelets. 2021. https://www.mathworks.com/help/wavelet/ug/lifting-method-for-constructing-wavelets.html.

[21] Okamoto, N., Yoshimatsu, K., Schneider, K., Farge, M. & Kaneda, Y., Coherent vortices in high resolution direct numerical simulation of homogeneous isotropic turbulence: A wavelet viewpoint. *Physics of Fluids*, 2007.

[22] Hastings, E.J., Mesit, J. & Guha, R.K., Optimization of large-scale, real-time simulations by spatial hashing. *Summer Simulation Conference*, 2005.

[23] Mallat, S.G., Multifrequency channel decompositions of images and wavelet models. *IEEE Transactions on Acoustics, Speech, and Signal Processing*, 1989.

[24] Sweldens, W., The lifting scheme: A custom-design construction of biorthogonal wavelets. *Applied and Computational Harmonic Analysis*, 1996.

MESH DISPERSION MINIMIZATION ALGORITHMS WITHIN EXPLICIT FINITE-DIFFERENCE SCHEMES TO CALCULATE TRANSIENT WAVE PROCESSES IN ELASTIC MEDIA AND COMPOSITE STRUCTURES

SAGDULLA ABDUKADIROV[1,2]
[1]Tashkent Institute of Architecture and Civil Engineering, Uzbekistan
[2]Tashkent Institute of Irrigation and Agricultural Mechanization Engineers, Uzbekistan

ABSTRACT

Precise calculation of wave fronts and high-gradient components is always of utmost importance for problems of numerical simulation of wave processes in media and composite structures. The usage mesh algorithms come across specific obstacles, which do not allow to accurately calculate such disturbances localized at the loading area or propagated with time. One of such obstacles (notably in the problems with singularities) is the spurious effect caused by the mesh dispersion responsible for the emergence of high-frequency "parasite" oscillations damaged the computer solution. In this work, advanced numerical algorithms within the explicit finite-difference scheme are developed exactly for very purpose – to precisely calculate wave processes with singularities. The algorithms are constructed with the condition that dependence domains are the same (or maximally closed) in differential and difference equations corresponding to continual and discrete models, respectively. In the designed algorithms, the influence of spurious effects of numerical dispersion is suppressed (or essentially minimized) that allows discontinuities in fronts and high-gradient components to be accurately calculated. A set of examples of computer simulations of linear and nonlinear wave processes are presented. Among them are (a) impact propagation in a waveguide resting on an elastic foundation, (b) cylindrical and spherical waves, and (c) wave propagation and fracture pattern in a unidirectional composite. Comparison of results calculated by conventional and developed algorithms clearly shows the advantage of the latter. To this end, precise numerical solutions (in mesh points of the discrete space) are obtained for the problems listed above.
Keywords: transient wave dynamics, explicit finite difference scheme, mesh dispersion, fracture of unidirectional composites.

1 INTRODUCTION

Significant rise in influence of the microstructure on the wave pattern essentially restricts capabilities of analytical modeling. On the other hand, numerical solutions allow to obtain quantitative evaluations of the process under study and to explain physical consequences. At the same time, mesh algorithms used in computer codes come across specific obstacles, which do not allow to calculate accurately wave fronts and high-gradient components localized at the loading area or propagated with time. One of such obstacles is the spurious effect caused by the *Mesh Dispersion* (*MD*) and responsible for the emergence of the high-frequency "parasite" noise damaged the computer solution. This phenomenon manifested notably in the problems with singularities and multiple reflections, possesses own high-frequency patterns which are typical for compound media and composites. The studies of the *MD* in initial-boundary problems have a long-standing history and extensive literature beginning with classical works [1]–[4]. The present study is the further development of the so-called *Mesh Dispersion Minimization* (*MDM*) procedure in the explicit finite difference algorithms originally presented in [5] and then used in calculations of diverse range of practical engineering problems in [6]–[11].

The *MDM* technique is based on a generalized concept of the Courant condition linked temporal and spatial mesh steps with the wave velocity, which reflects properties of the material at hand. An important technical advantage of *MDM* is that it utilizes the same homogeneous mesh for both high-gradient and smoothed wave components.

The *MDM* algorithm for dispersionless waveguides proves to be stable at the Courant number $\lambda = c_0 \Delta t / \Delta x = 1$ [5], where Δt and Δx are the time and spatial dimensions of the difference mesh, c_0 is the sound velocity. The equality $\lambda = 1$ has the simple physical sense: during one step in time, the wave passes one spatial step. Here the solutions of continual and difference problems coincide in mesh points.

The algorithms controlling *MD* have been also developed in many contemporary studies, see for example, [12]–[17] intended for a wide spectrum of physical problems; among them the application to the analysis of even financial and stock exchange processes is found (see [12]). Unfortunately, one of important points of *MD* suppression – the accuracy of computation of wave processes possessing front gaps is not still completely revealed.

In this work, we have built algorithms suppressed *MD* in linear and nonlinear dispersive waveguides (Section 2). In Section 3, cylindrical and spherical waves possessing fronts are precisely calculated. Finally, in Section 4, we have presented *MDM* algorithms and examples of computer simulations of the multi-front wave pattern and development of the dynamic fracture realized in a unidirectional composite plate.

2 ELASTIC WAVEGUIDES

2.1 Classical dispersionless waveguide

Consider the wave propagation problem in a semi-infinite dispersionless waveguide ($x \geq 0$) subjected by the tension force F at free end $x = 0$. We have used the model:

$$\ddot{u} = c_0^2 u'', \quad u'(0,t) = -F(t), \quad u(x,0) = \dot{u}(x,0) = 0, \tag{1}$$

where $u = u(x,t)$ is displacement, c_0 is the sound speed, while parameters of the waveguide serves as measurement units. The goal is to choose the most effective calculation scheme.

This simple model (1) is very convenient for demonstration of the main *MDM* principle.

First of all, we apply the Fourier expansion $u = U \exp[\iota q(ct \pm x)]$ to eqn (1) and obtain the dispersion relation $c = c_0$ inherent to the dispersionless free wave propagation in the infinite waveguide ($|x| < \infty$). Above indicated: $U \sim const$, ι is the imaginary unit, q is the wave number (then $l = 2\pi/q$ is the wavelength), c is the phase velocity.

The explicit "cross" type finite-difference analog of eqn (1) is written as follows:

$$u_i^{k+1} = 2u_i^k - u_i^{k-1} + \lambda^2 (u_{i+1}^k - 2u_i^k + u_{i-1}^k), \quad \lambda = c_0 \Delta t / \Delta x, \tag{2}$$

where $x = i\Delta x$ ($i = 0, \pm 1, \pm 2, \ldots,$), $t = k\Delta t$ ($k = 0, 1, 2, \ldots$), i and k are coordinates of the mesh nodes, λ is the Courant number.

Substituting Fourier expansion $u_i^k = U \exp[\imath q(ck\Delta t \pm i\Delta x)]$ into eqn (2), we have obtained the following dispersion relation in form $c = c$ (q): $c = (2/q\Delta t)\arcsin[\lambda \cdot \sin(q\Delta x/2)]$ proved the dependence of the phase velocity on the wavenumber q and mesh steps that, generally speaking, determines the wave dispersion in model (2).

At the same time, however, one can detect that in the case $\lambda = 1$ this dispersion relation becomes the same that in the continual case: $c = c_0$, and the MD is completely eliminated.

In a general case, we have no closed analytical solution of eqn (2) like that for eqn (1) with added initial and boundary conditions, but in case $\lambda = 1$ such a solution is obtained with using the mathematical induction technique (see [5]).

Let us compare solutions of eqns (1) and (2) for two versions of boundary loadings at the end $x = 0$: (a) the Heaviside step tension $F(t) = -H(t)$ and (b) the Dirac pulse $F(t) = -\delta(t)$; c_0 serves as measurement unit:

$$\text{eqn (1):} \quad (a)\ u'(x,t) = -H(t-x); \quad (b)\ u'(x,t) = -1\ (x = t),\ u'(x,t) = 0\ (x \neq t),$$
$$\text{eqn (2):} \quad (a)\ u'(x,t) = -H(k-i); \quad (b)\ u'(x,t) = -1\ (k = i),\ u'(x,t) = 0\ (k \neq i). \tag{3}$$

(for eqn (1), the well-known d'Alambert solution is used). These solutions coincide in mesh nodes. Such a computer solution is defined as "*the accurate numerical solution*" (ANS). Snapshots of strains u' at $t = 100$ are shown in Fig. 1. Note, computer results obtained with $\lambda < 1$ are essentially distorted by the MD action.

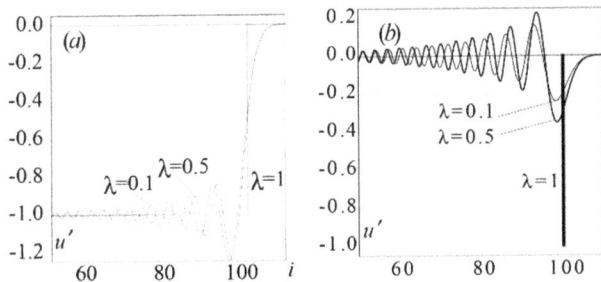

Figure 1: Snapshots of strains at $t = 100$ for two kinds of loadings. (a)The Heaviside step; and (b) The Dirac pulse. Oscillating curves are numerical solutions at $\lambda < 1$.

2.2 Dispersive waveguides

Consider now wave equation

$$\ddot{u} - c_0^2 u'' + G(u) = 0 \tag{4}$$

known as the generalized Klein–Gordon equation, where $G(u)$ is a finite function. Eqn (4) is widely used in diverse models of physics and structure dynamics.

First, consider linear function $G(u) \equiv gu$, $g \sim const$. Then eqn (4) will be written as

$$\ddot{u} - c_0^2 u'' + gu = 0 , \tag{5}$$

and its dispersion equation is

$$c(q) = \sqrt{1 + g / q^2} , \tag{6}$$

which, in contrast with relation $c = q$ corresponding to eqn (1) shows the wave dispersion. The explicit "cross"-type finite-difference analog of eqn (1) is

$$u_i^{k+1} = 2u_i^k - u_i^{k-1} + \lambda^2 (u_{i+1}^k - 2u_i^k + u_{i-1}^k) + (\Delta t)^2 gu_i^k \quad (\lambda = \Delta t / \Delta x), \tag{7}$$

where Δx and c_0 are taken as measurement units: $\Delta x = c_0 = 1$, that results in equality $\lambda = \Delta t$. The dispersion relation for eqn (7) is obtained as

$$c = \pm (2/q\Delta t) \arcsin \left(\Delta t \sqrt{\sin^2 (q/2) + g/4} \right) , \tag{8}$$

that proves presence of the *MD* inevitably distorted the computer solution.

Our aim is to construct such a difference scheme for the discrete analog of eqn (5) so that its dispersion relation would be as close as possible to relation (6). The *MDM* approach helps to achieve this goal. We have changed the ordinary local approximation $u = u_i^k$ in term gu of eqn (7) to a special non-local three-point approximation:

$$u \sim (u_{i+1}^k + 2u_i^k + u_{i-1}^k) / 4 . \tag{9}$$

Then, the conventional algorithm (7) is turned to be written as the following:

$$u_i^{k+1} = 2u_i^k - u_i^{k-1} + \lambda^2 [(u_{i+1}^k - 2u_i^k + u_{i-1}^k) + g(u_{i+1}^k + 2u_i^k + u_{i-1}^k) / 4] . \tag{10}$$

(there can be readily shown that approximation orders of eqns (7) and (10) are the same $- (\Delta x)^2 + (\Delta t)^2$, while the dispersion relation for eqn (10) acquires the following form:

$$c = \pm (2/q\Delta t) \arcsin \left[\Delta t \sqrt{\sin^2 (q/2) + (g/4) \cos^2 (q/2)} \right] . \tag{11}$$

If we set $\lambda = 1$ (then $\Delta t = 1$ due to the fact that $\Delta x = c_0 = 1$) in eqn (10), then extremely short waves of length $l = 2$ ($q = \pi$) will propagate with the same phase velocity $c = 1$ as infinitely short waves, $l \to 0$ ($q \to \infty$), in the continual model. As in the free waveguide above, *MD* is completely eliminated over the entire discrete spectrum.

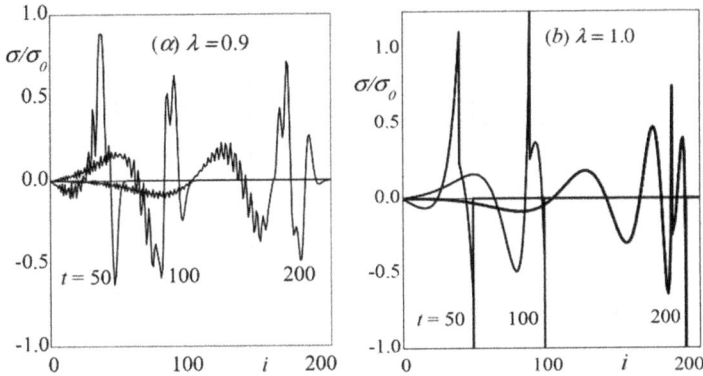

Figure 2:   Computer solutions for normalized stresses propagated along the waveguide ($g = 0.01$) under action of pulse of duration $t_0 = 10$. (a) Conventional algorithm (7) with $\lambda = 0.9$. (b) MDM algorithm (10) with $\lambda = 1$ resulting in the ANS.

In Fig. 2, results of computation of wave propagation processes are compared. We have added zero initial conditions to eqn (10) and the boundary condition as action of the step-wise compression stress at the end $x = 0 : \sigma(0, t) = -\sigma_0 H(t - t_0)$, where σ_0 and t_0 are the pulse amplitude and duration.

The essential distortions are detected of the solution obtained by the conventional algorithm (7) with $\lambda = 0.9$, while calculations with the MDM algorithm (10) with $\lambda = 1$ can be considered as the AMS: spurious oscillations are absent, and front gaps are clearly detected.

Calculation results below are obtained with MDM algorithms.

The introduced above three-point approximation (9) within MDM-algorithms used also in the case of the inhomogeneous foundation: $G(u) = g(x)u(x)$. Although the dispersion equation is absent here, using the so-called method of frozen coefficients helps achieve our goal: we denote $g = \max \left| g(x) \right|_x$ and change variables in eqn (4): $\bar{x} = x\sqrt{g}$, $\bar{t} = t\sqrt{g}$.

After that MDM algorithm (10) is launched.

Consider now longitudinal wave propagation processes in a semi-infinite waveguide ($x \geq 0$) resting upon a nonlinear foundation. Let boundary and initial conditions be the same that were used above. In the example below we consider the following initial-boundary problems:

$$\ddot{u} - c_0^2 u'' + gu(1 + g_0 u^2) = 0, \quad \varepsilon(0, t) \equiv u'(0, t) = -F(t), \quad u(x, 0) = \dot{u}(x, 0) = 0. \quad (12)$$

Our aim is to design the MDM algorithm to problem (12) and to reveal by computer simulations the influence of nonlinearity on the wave propagation process.

Completed tests show that the MDM approximation (9) together with condition $\lambda = 1$ results in the dispersionless algorithm

$$u_i^{k+1} = 2u_i^k - u_i^{k-1} + \lambda^2 [(u_{i+1}^k - 2u_i^k + u_{i-1}^k) + gU(1 + g_0 U^2)], \quad U = (u_{i+1}^k + 2u_i^k + u_{i-1}^k)/4, \quad (13)$$

allowing the problem on the same mesh and with the same accuracy to be calculated as in the linear case.

In Fig. 3, snapshots of linear and nonlinear stress wave patterns are compared at $t = 250$. Note, the fundamental difference in front zone is not found in linear and nonlinear solutions (despite the relatively huge value of g_0). This surprising (at first sight) result can be explained by the fact that the package of high frequency oscillations generating in the front zone, propagates together with the front, while the perturbations related to the presence of the foundation (and, in this way, the nonlinearity) moves behind the front zone.

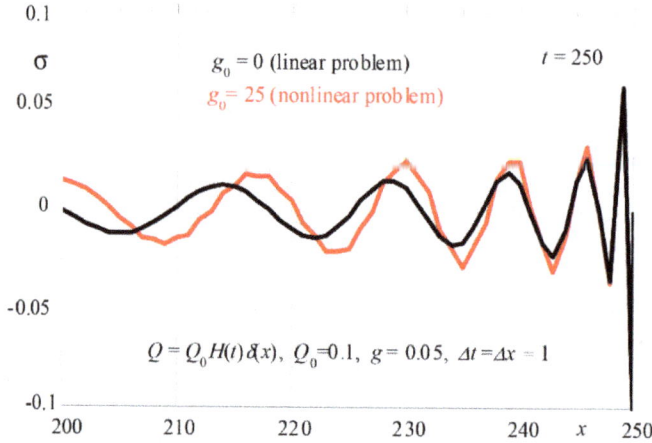

Figure 3: Snapshots (at $t = 250$) of the strain distributions in linear and nonlinear problems.

3 SPHERICAL AND CYLINDRICAL WAVES

Consider the following wave equation possessing the inhomogeneous term:

$$\ddot{u} = c_0^2 u'' + B(x)u' , \tag{14}$$

where $B(x)$ is the finite function which can describe various models of continual media, for example, elastic waves in a thin rod of the variable the cross-section area or cylindrical and spherical waves in a compressible liquid under action of linear or point sources, respectively. In general, eqn (14) has no dispersion relation. As above, in Section 2, we apply the method of frozen coefficients (assume $|B(x)| \leq \bar{B} \sim const$) and then use the standard Fourier analysis to the difference analog of modified eqn (14). Then we obtained:

$$u_i^{k+1} - 2u_i^k + u_i^{k-1} = \lambda^2 (u_{i+1}^k - 2u_i^k + u_{i-1}^k) + \lambda \Delta t \, \bar{B}(u_{i+1}^k - u_{i-1}^k)/2 , \ \lambda = \Delta t / \Delta x . \tag{15}$$

The resulting dispersion equation,

$$c(q, \Delta x, \Delta t) = \frac{2}{q\Delta t} \arcsin\left[\lambda \sqrt{\sin^2 \frac{q\Delta x}{2} + \frac{\bar{B}\Delta x}{4} \sin(q\Delta x)} \right] , \tag{16}$$

shows that if to put $\lambda = 1$ (remind, c_0 and Δx are measurement units, then $\Delta t = 1$), then. maximally short waves $(q = \pi)$ will propagate with phase velocity $c = 1$. Now the *MDM* algorithm for calculation of eqn (15) with equality $\lambda = 1$ is to be as follows:

$$u_i^{k+1} = u_{i+1}^k + u_{i-1}^k - u_i^{k-1} + \overline{B}(u_{i+1}^k - u_{i-1}^k)/2 \, . \tag{17}$$

Consider cylindrical and spherical waves in a compressible liquid and set $B(x) = \alpha/x$. Then eqn (17) rewritten as

$$\ddot{u} = u'' + (\alpha/x)u', \ x > 0$$

will describe cylindrical ($\alpha = 1$) and spherical ($\alpha = 2$) pressure waves. Here u plays role of the velocity potential, x is the radial coordinate. In the wave, radial velocity and pressure are expressed as $\dot{u} = \partial u/\partial x$ and $P = -\partial u/\partial t$, respectively (the bulk compression modulus and the liquid density serve as measurement units). Under the condition $\lambda = 1$, eqn (17) is turned out to the following *MDM* algorithm (here $\Delta x = \Delta t = c_0 = 1$):

$$u_i^{k+1} = u_{i+1}^k - u_i^{k-1} + u_{i-1}^k + \frac{\alpha}{2(x_0 + i)}(u_{i+1}^k - u_{i-1}^k), \tag{18}$$

where x_0 is the source radius, measured in steps Δx.

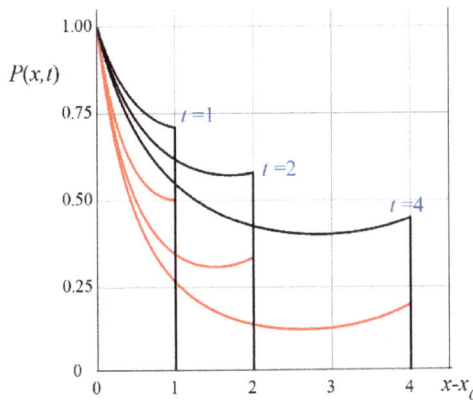

Figure 4: Pressure distribution at the moments of time $t = 1, 2, 4$ in spherical (red lines) and cylindrical (black lines) waves under action of pressure $P(1,t) = H(t)$.

Let pressure of the Heaviside type, $P(1,t) = H(t)$, be set on cavity surfaces ($x_0 = 1$) in both cases. In Fig. 4, calculated by *MDM* algorithm (18), snapshots are shown of the pressure distributions along the radial coordinate at time values $t = 1, 2$ and 4. The obtained numerical solutions completely agree with analytical ones (see, e.g. [18]). So, the *ANS*s are presented here.

Note that an analytical solution for the cylindrical case is not represented in a closed form and can be available as asymptotic one only in the small vicinity of the front.

4 FRACTURE DYNAMICS OF FIBER REINFORCED COMPOSITE PLATE

The process of stresses concentration and stepwise crack propagation at fiber-adhesive interfaces of the fiber reinforced composite plate is numerically simulated. Note, some calculation results of delamination phenomena in layered structures were presented in [19], but the stepwise crack phenomena was not considered.

The problem statement corresponds to the following description: the material of the plate is stretched along the fibers at infinity by a constant tensile stress σ_∞ (see Fig. 5). Here E and G are the Young module in fibers (black) and the shear module in adhesive (grey), ρ and ρ_a are densities, h and H are thicknesses, respectively. At zero time ($t = 0$), one of the fibers (say it be fiber $m = 0$) starts to fail due to some defects.

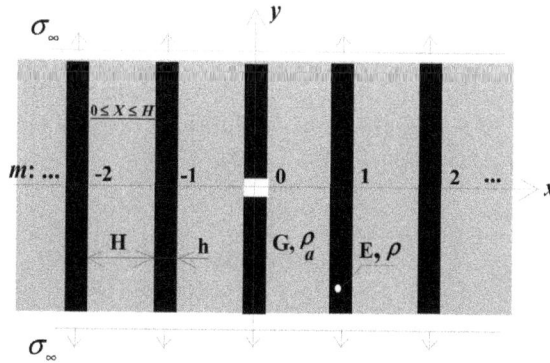

Figure 5: Loading of a unidirectional composite with the damage in a fiber.

Let E and G be the Young modulus in fibers and the shear module in adhesive respectively, ρ and ρ_a be densities, h and H be thicknesses of components as shown in Fig. 5, $c_f = \sqrt{E/\rho}$ and $c_a = \sqrt{G/\rho_a}$ be sound speeds in fibers and adhesive, respectively. Fiber constants E, ρ, h serve as measurement units.

The fractured fiber starts to unload, and the intact ones start to load up due to action of shear stress waves propagating in the adhesive. Along with this process, the delamination events can happen depending on the strength of the adhesive.

The used model of the fiber dynamics describes the one-dimensional wave process in a thin rod embedded into the adhesive, which represented as inertial bonds perceived shear stresses (tension–compression stresses in adhesive bonds are neglected). Such a theoretical treatment of the components performance can be justified by the fact that the shear modulus of adhesive is much less than that of fiber (see e.g. [20]), while their stretches have roughly the same level due to the cohesion of fibers and the adhesive. When maximal stresses reached in adhesive do not exceed the strength limits ($\tau_m < \tau^*$), the adhesive remains intact. The crack propagation is investigated on the basis of linear elastic fracture mechanics: fracture in adhesive is initiated if $\tau_m \geq \tau^*$, and then propagated deep into the composite up to their stop due to continuous scattering of the initial impact energy with time.

4.1 Mathematical formulation

In the mathematical sense, we met a non-linear hyperbolic problem possessing non-classical boundary conditions. Due to the natural symmetry, a quarter of plane x, y is considered in the calculation algorithm (let it be $x \geq 0$, $y \geq 0$). At the initial state ($t < 0$) displacements and strains in fiber are

$$u_m(y) = y \sigma_\infty / E, \text{ or } \varepsilon_m(y) = \sigma_\infty / E; \; v_m(X, y) = 0 \; (m = 0, \pm 1, \pm 2, \ldots), \quad (19)$$

where $\varepsilon_m(y) = \partial u_m / \partial y$ is the strain in mth fiber ($m \neq 0$), and the fracture event of fiber $m = 0$ at $t = 0$ changes eqn (19) by adding condition $\varepsilon_0(0, t) = 0$:

$$\varepsilon_0(0) = 0; \; \varepsilon_m(y) = \sigma_\infty / E \; \left(m \neq 0, \; y \neq 0 \right), \; v_m(X, y) = 0 \; (m = 0, \pm 1, \pm 2, \ldots). \quad (20)$$

Then reformulate the problem for the additional dynamic state subtracting the static strains (19) from eqn (20). Then boundary conditions for strains in fibers are the following:

$$y = 0: \; \varepsilon_0(0, t) = \left(\partial u / \partial y \right)_{m=0} = -\sigma_\infty / E, \; \varepsilon_m(0, t) = \left(\partial u / \partial y \right)_{m \neq 0} = 0. \quad (21)$$

The motion of fibers is described by the system of *1D* wave equations

$$\rho h \ddot{u}_m = E h u''_{m, y} + \tau_m^+(y) - \tau_m^-(y), \; m = 0, \pm 1, \pm 2, \ldots, \quad (22)$$

where τ_m^+ and τ_m^- correspond to reactive shear forces at the fiber-adhesive interface on the right and the left, respectively:

$$\tau_0^+ = -\tau_0^- = G v'_{0,X} \big|_{X=0}, \; \tau_m^+ = G v'_{m,X} \big|_{X=0}, \; \tau_m^+ = G v'_{m-1,X} \big|_{X=H} \; (m > 0), \quad (23)$$

while displacements in adhesive described by wave equations

$$\partial^2 v_m(X, y, t) / \partial t^2 = c_a^2 \, \partial^2 v_m(X, y, t) / \partial X^2 \; (0 \leq X \leq H), \, m = 0, 1, 2, \ldots \quad (24)$$

with boundary conditions:

$$v_m(0, y, t) = u_m(y, t), \; v_m(H, y, t) = u_{m+1}(y, t). \quad (25)$$

Then the following additional relations in expressions of reactive forces (23) and boundary conditions (25) are:

$$\tau_m^\pm \left(\xi^\pm, y, t \right) \geq \tau^* \Rightarrow t^*_{m,\pm} = t; \; t > t^*_{m,\pm}: \; \tau_m^\pm \left(\xi^\pm, y, t \right) = 0, \; \partial v_m^\pm \left(\xi^\pm, y, t \right) / \partial X = 0, \quad (26)$$

where $\xi^+ = 0$, $\xi^- = H$, while indices "$\pm$" at t_m^* denote right and left interfaces, respectively.

4.2 The *MDM* calculation algorithm

It is evident that each possible scenario of wave-fracture pattern is saturated by reflected waves with discontinuities appeared due to adhesive cracking. Our goal is to calculate such processes as precisely as possible. Below we present the practical calculation device based on the *MDM* technique allowing this goal to be reached. Let the mesh step in adhesive be Δx. Then the *MDM* conditions are $\Delta t = \Delta y = 1$, $\Delta x = c_a$, and the adhesive dynamics, as the analogue of eqn (24), described by the following *MDM*-algorithm:

$$v_{m,j,i}^{k+1} = v_{m,j,j+1}^{k+1} + v_{m,j,i-1}^{k} - v_{m,j,i}^{k-1} \quad (0 \le i \le s = H/\Delta x), \tag{27}$$

while the *MDM*-algorithm of fibers dynamics in the difference analogue of eqn (22) can be written as

$$u_{m,j}^{k+1} = u_{m,j+1}^{k} + u_{m,j-1}^{k} - u_{m,j}^{k-1} + \kappa F, \quad F = \left(v_{m,j,i}^{k} + v_{m-1,j,i}^{k}\right) \ (m > 0), \quad F = 2v_{1,j,i}^{k} \ (m = 0),$$
$$\kappa = GH/(Eh\mu), \quad \mu = h\rho_f \Delta y + H\rho_a \Delta x. \tag{28}$$

4.3 Results of computer simulations

In calculation examples below, parameters of composite are: $H = 5$, $G = 0.025$, $\rho_a = 0.4$; stresses are normalized to σ_∞, critical values "σ_* and "τ_* are varied.

Let us turn to examples. The patterns of normal stresses σ_m in fibers $m = 1$, 2 and 3 vs. time and shear stresses in interfaces $X = 0$ and $X = H$ of the adhesive layer at $y = 0$ are shown in Fig. 6.

It is convenient to relate stresses to their asymptotic values $\tau_{st} \sim 0.053$, $\sigma_{st} = 4/3$ $(t \to \infty)$ obtained in [21]. We note that peak amplitudes of shear stresses playing the main role in the fracture initiation process can be much more than τ_s. The peaks are appeared with the period $t = 2H/c_a$ due to reflections of shear waves, and their values do not change with time.

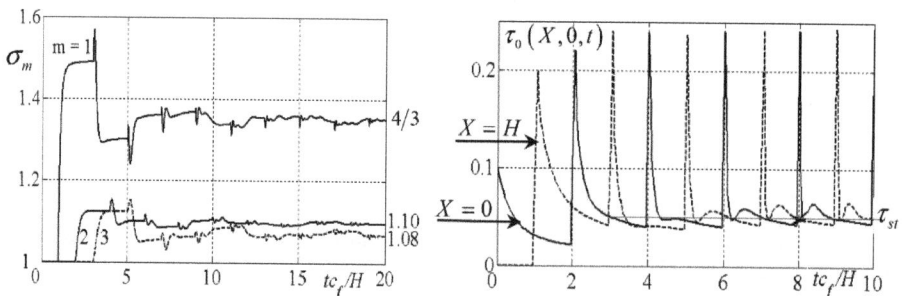

Figure 6: Normal stresses in fibers, σ_m, and shear stresses in adhesive layer, $m = 0$.

The fracture propagation pattern in the case $\sigma^* = 1.2$ and $\tau^* = 0.1$ is shown in Fig. 7. After the initial rupture of 0th fiber ($t = 0$), the fracture in the adhesive occurs at the fiber-adhesive interfaces $X = 0$ (red), $X = H$ (blue) and possesses a high-speed avalanche-like pattern; the speed of fracture propagation decreases with time and the adhesive fracture is stopped at $t = 30$ in interface $X = 0$. Adhesive layers with $m \neq 0$ remain intact. Rapture of fibers occurs at $y = 0$ and propagates with a little change in speed up to $t = 63 H/c_f$, then it stopped.

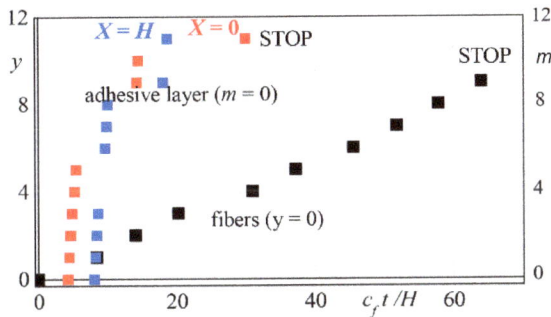

Figure 7: Fracture pattern in the composite vs. time ($\sigma^* = 1.2$, $\tau^* = 0.1$).

Obtained results can be summarized in the main conclusions:

1. Designed *MDM*-algorithms allows high-gradient and step-wise character of wave processes to be precisely calculated.
2. The accurate *MDM*-solutions described front propagation patterns in media and structures are obtained for a set of linear and non-linear *1D* transient wave problems.
3. The *MDM*-algorithms built for exploring the *2D* processes of stress concentrations and progressive fracture of fiber reinforced composite plates allow one to analyze fracture development heavily caused by discontinuities of shear waves propagated in the adhesive.

REFERENCES

[1]   Weinberger, H.F., Upper and lower bounds for eigenvalues by finite difference methods. *Communications on Pure and Applied Mathematics*, **9**, pp. 613–623, 1956.
[2]   Lax, P.D. & Wendroff, B., Difference schemes for hyperbolic equations with high order of accuracy. *Communications on Pure and Applied Mathematics*, **17**, pp. 381–392, 1964.
[3]   Fromm, J.E., A method for reducing dispersion in convective difference schemes. *Journal of Computational Physics*, **3**, pp. 176–183, 1968.
[4]   Chin, R.C.Y., Dispersion and Gibbs phenomenon associated with difference approximations to initial boundary-value problems for hyperbolic equations. *Journal of Computational Physics*, **18**(3), pp. 233–247, 1975.
[5]   Abdukadirov, S.A., Pinchukova, N.I. & Stepanenko, M.V., A numerical solving dynamic equations of elastic media and structures. *Journal of Mining Science*, **6**, pp. 19–22, 1984.

[6]   Belov, A.I., Kornilo, V.A. & Stepanenko, M.V., Reaction of a three-layer hydroelastic cylindrical shell on the effect of axisymmetric internal explosion. *Journal of Applied Mechanics and Technical Physics*, **1**, pp. 49–57, 1986.

[7]   Abdukadirov, S.A., Kurmanaliev, K.K. & Stepanenko, M.V., Dynamic stresses on the perimeter of a rigid inclusions embedded in a rock bed. *Journal of Mining Science*, **24**, pp. 442–447, 1988.

[8]   Abdukadirov, S.A., Alexandrova, N.I. & Stepanenko, M.V., Non-steady diffraction of a plane longitudinal wave on an elastic cylindrical shell. *Mechanics of Solids*, **5**, pp. 132–137, 1989.

[9]   Slepyan, L.I. & Ayzenberg-Stepanenko, M.V., Penetration of metal-fabric composite targets by small projectiles. *Personal Armor Systems*. British Crown Copyright/MOD: Colchester, UK, 1998.

[10] Ayzenberg-Stepanenko, M.V. & Slepyan, L.I., Localization of strain and melting wave in high-speed penetration. *IUTAM Symposium Nonlinear Singularities*, Kluwer, 1999.

[11] Kubenko, V.D. & Ayzenberg-Stepanenko, M.V., Impact indentation of a rigid body into elastic layer. *Journal of Mathematical Sciences*, **1**, pp. 156–171, 2009.

[12] Int-Veen, R., Avoiding numerical dispersion in option valuation. *Computer Visual Science*, **10**(4), pp. 556–564, 2007.

[13] Wu, Y.S. & Forsyth, P.A., Efficient schemes for reducing numerical dispersion in modeling multi-phase transport through heterogeneous geological media. *Vadose Zone Journal*, **7**(1), pp. 340–349, 2008.

[14] Sun, Z., Ren, Y., Larricq, C., Zhang, S. & Yang, Y., A class of finite difference schemes with low dispersion and controllable dissipation for DNS of compressible turbulence. *Journal of Computational Physics*, **230**(12), pp. 231–244, 2011.

[15] Zhou, Y., Yang, D., Ma, X. & Li, J., An effective method to suppress numerical dispersion in 2D acoustic and elastic modeling using a high-order Padé approximation. *Journal of Geophysics and Engineering*, **12**(1), pp. 114–129, 2015.

[16] Wu, Z. & Alkhalifah, T.A., Highly accurate finite-difference method with minimum dispersion error for solving the Helmholtz equation. *Journal of Computational Physics*, **365**, pp. 350–361, 2018.

[17] Cheng, Y., Guangzhi, C., Xiang-Hua, W. & Shunchuan, Y., Investigation of numerical dispersion with time step of the FDTD methods: avoiding erroneous conclusions. *IET Microwaves, Antennas and Propagation*, pp. 1–13, 16 Apr. 2021. DOI: 10.1049/mia2.12068.

[18] Slepyan, L.I., *Nonstationary Elastic Waves*, Sudostrojenie: Leningrad, 1972. (In Russian.)

[19] Nayak, A.K., Shenoi, R.A. & Moy, S.J., Transient response of initially stressed composite plates. *Finite Elements in Analysis and Design*, **2**(10), pp. 821–836, 2006.

[20] Summerscales, J. ed., *Microstructural Characterisation of Fibre-Reinforced Composites*, Woodhead Publishing: Cambridge, 1998.

[21] Michailov, A.M., Dynamics of a unidirectional fiberglass. *Journal of Applied Mechanics and Technical Physics*, **4**, pp. 139–145, 1974.

SECTION 2
MATERIALS
CHARACTERIZATION

MICROSTRUCTURE AND FATIGUE PROPERTIES OF CU–NI–SI ALLOY STRENGTHENED BY NI₂SI INTERMETALLIC COMPOUNDS

MASAHIRO GOTO[1], TAKAHITO UTSUNOMIYA[1], TAKAEI YAMAMOTO[1], SEUNG ZEON HAN[2],
JUNICHI KITAMURA[1], JEE-HYUK AHN[2], SUNG HWAN LIM[3] & TERUTOSHI YAKUSHIJI[4]
[1]Department of Mechanical Engineering, Oita University, Japan
[2]Korea Institute of Materials Science, Changwon, Republic of Korea
[3]Department of Advanced Materials Science and Engineering, Kangwon National University, Republic of Korea
[4]National Institute of Technology, Oita College, Japan

ABSTRACT

Microstructure and fatigue properties of Cu–6Ni–1.5Si alloy having different morphologies of Ni₂Si intermetallic compounds that are disk-shaped continuous precipitates (CPs) with nano-size diameter by normal aging, fibre-shaped discontinuous precipitates (DPs) by overaging and elongated DPs fabricated by cold-rolling (DPR) were studied. There was a negligible difference in fatigue strength between the CP and DP specimens despite higher tensile strength of the CP specimen. The DPR specimen had the highest tensile and fatigue strengths in all specimens. The fatigue crack initiation resistance of the DPR specimen was drastically enhanced. The growth rate of a small crack can be determined by a term $\sigma_a{}^d l$. The crack growth resistance of the DPR specimen was nearly equal to that of the CP specimen. The reason for such trends of tensile and fatigue strengths was discussed based on the microstructure of each specimen.

Keywords: fatigue, copper alloy, crack initiation, microstructure, cold-rolling.

1 INTRODUCTION

Cu–Ni–Si alloys [1] used for lead frame, connector applications are commercially manufactured through normal aging after solution heat treatment (SHT). Normal aging of Cu–6Ni–1.5Si alloy produced continuous precipitates (CPs) of nano-sized, δ-Ni₂Si intermetallic compounds which brought a high tensile strength throughout the matrix [1]–[4]. It is well known that an enhancement in the tensile strength of normal aged Cu alloys is inevitably accompanied by a reduction in electrical conductivity. On the other hand, overaging of Cu–Ni–Si alloys, particularly with high solute concentrations, produced discontinuous precipitates (DPs) that were fibre-shaped, stable δ-Ni₂Si intermetallic compounds in the Cu matrix. Mechanical properties of Cu–Ni–Si alloys are degraded by the formation of DPs [5]–[9], however DPs have a superior electrical conductivity. It has been shown that the DP specimen subjected to cold working had elongated nanofiber-shaped δ-Ni₂Si precipitates and the tensile strength was enhanced without a significant loss of electrical conductivity [8].

For the envisaged structural applications, fatigue strength should be clarified, because that at least 90% of mechanical failures during service were caused by the fatigue failure. To make the design and maintenance of safe machine components and structures, the fatigue behaviour such as crack initiation and propagation should be clarified. Up to now, there was a small number of studies [10]–[16] on fatigue behaviour of Cu–Ni–Si alloys, particularly in DPs. In the present study, fatigue tests of Cu–6Ni–1.5Si alloy with different morphologies of Ni₂Si precipitates were carried out at the room temperature. The objective of this study is to investigate the microstructure and fatigue crack growth behaviour of Cu–Ni–Si alloy with different morphologies of Ni₂Si precipitates.

WIT Transactions on Engineering Sciences, Vol 130, © 2021 WIT Press
www.witpress.com, ISSN 1743-3533 (on-line)
doi:10.2495/CMEM210071

For the envisaged structural applications, fatigue strength should be clarified, because that at least 90% of mechanical failures during service were caused by the fatigue failure. To make the design and maintenance of safe machine components and structures, the fatigue behaviour such as crack initiation and propagation should be clarified. Up to now, there was a small number of studies [10]–[16] on fatigue behaviour of Cu–Ni–Si alloys, particularly in DPs. In the present study, fatigue tests of Cu–6Ni–1.5Si alloy with different morphologies of Ni_2Si precipitates were carried out at the room temperature. The objective of this study is to investigate the microstructure and fatigue crack growth behaviour of Cu–Ni–Si alloy with different morphologies of Ni_2Si precipitates.

2 EXPERIMENTAL PROCEDURES

Using 99.9% pure Ni, and 99.99% pure Si as alloying elements, Cu–6wt%Ni–1.5wt%Si alloy was cast by induction melting. The cast was cold-rolled with 80% reduction in area, and subsequently solution heat-treated at 980°C for 1 h with water quenching. The solution heat-treated bars were aged at 500°C. The aging time was 0.5 and 3 h for CP and DP structure, respectively. In what follows, the specimen aged for 0.5 and 3 h is designated by the CP and DP specimen, respectively. Some of the cast were hot-rolled down to 75% reduction in area, and solution heat-treated at 980°C for 1 h. The solution heat-treated specimen was aged at 500°C for 3 h to obtain fully DP structure, then cold-rolled down to 80% reduction in area, which is designated as "DPR specimen".

The microstructure was observed using an optical microscope (OM) and a scanning electron microscope (SEM). A 200 kV field-emission transmission electron microscope (TEM) was utilized to characterize precipitates. Disks 3 mm in diameter and 100-μm-thick were prepared for the TEM observation by mechanical polishing with a digitally enhanced precision specimen grinder and dimpling by a dimple grinder. The microhardness (H_v) was measured using a Vickers hardness tester with an applied load of 1 N. Tensile tests (4 mm diameter specimens) were performed on a tensile testing machine (4206, Instron) with a strain rate of 0.017 s^{-1} at room temperature.

Fatigue specimens were round-bar with 5 mm diameter (Fig. 1) which machined from the CP, DP and DPR samples. To eliminate any surface damage induced during specimen preparation, the layer of approximately 25 μm of the specimen's surface was removed by electrolytic polishing prior to fatigue testing. All fatigue tests were conducted at room temperature using a rotating bending fatigue machine (constant bending-moment type, the stress ratio: $R = -1$) operating at 50 Hz. The observation of fatigue damage on the specimen surface and on the fracture surface was conducted using OM and SEM. The crack length, l, was measured along the circumferential direction of the surface using a plastic replication technique. The stress value referred to is that of the nominal stress amplitude, σ_a, at the minimum cross-section (5 mm diameter).

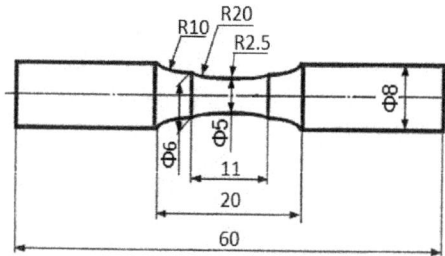

Figure 1: Schematic illustration of the fatigue specimen (in mm).

The electrical resistivity of a 300 mm length specimen was measured by a portable double bridge apparatus (2769, Yokogawa M&C) at room temperature ($27 \pm 1°C$). It was then converted to electrical conductivity by taking inverse of the resistance. The electrical conductivity was represented by the value of International Annealed Copper Standard (%IACS) which was calculated by the ratio of electrical conductivity between sample and annealed pure Cu.

3 EXPERIMENTAL RESULTS AND DISCUSSION

Fig. 2 shows OM micrograph of microstructure, Vickers hardness and electrical conductivity as a function of aging time. The hardness shapely increased with an increasing aging time. At 0.5 h aging, it reached near maximum value ($H_v = 259$), followed by a gradual decreasing trend. The 0.5 h aged microstructure consisted of bright gains and sporadically distributed dark/tarnished phases. The dark phases were DPs. After 3 h aging, the matrix was transformed to fully DP phases (DP fraction was over 95%) and hardness dropped to 196. The electrical conductivity (%IACS) gradually increased with an increase in aging time. The conductivity for 0.5 h aging was 25%IACS. After 3 h aging, it was drastically enhanced to 44%IACS. Even though the formation of DPs was detrimental to the mechanical properties of Cu–Ni–Si alloys, the electrical conductivity of DPs was superior to that of the CP counterpart. The increased conductivity of DPs was caused by the further reduction of Ni and Si solute elements in the Cu matrix.

3.1 Microstructure of the CP specimen

Fig. 3 shows the microstructure of the CP specimen. There were bright grains, dark/tarnished phases which sporadically distributed along grain boundaries (GBs) and inclusions with a

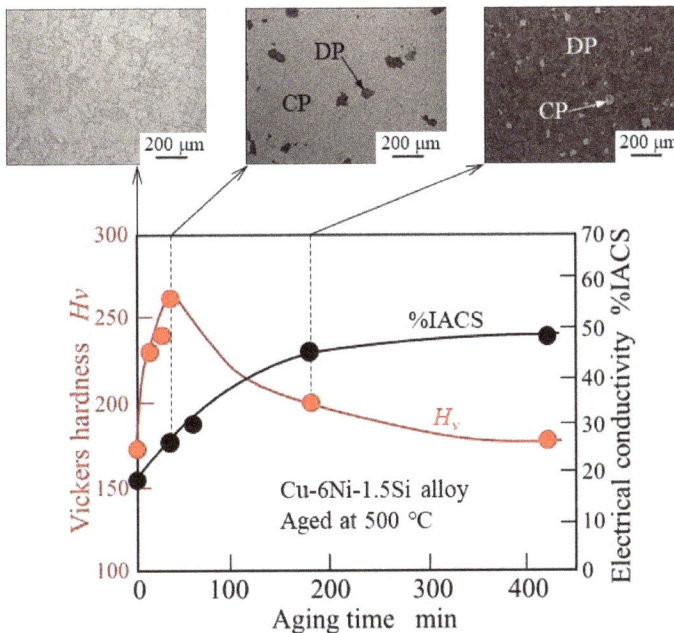

Figure 2:   OM micrograph of microstructure, Vickers hardness and electrical conductivity as a function of aging time.

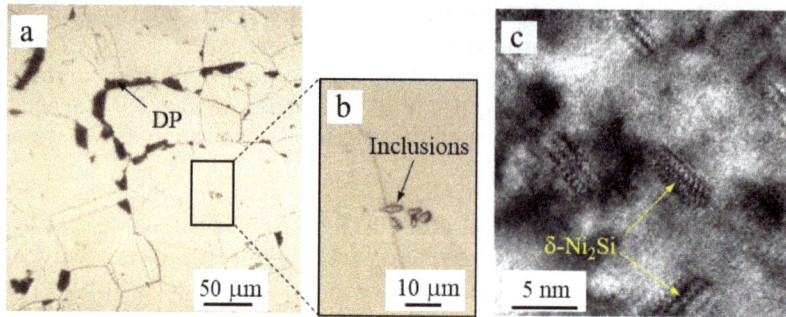

Figure 3:   Microstructure of the CP specimen. (a), (b) OM micrograph; and (c) high-resolution TEM image of the matrix.

few micrometres in size (Fig. 3(a) and (b)). Fig. 3(c) shows high resolution TEM images of the matrix (bright grains in Fig. 3(a)). Disc-shaped precipitates with a few nanometres in diameter were observed. These precipitates were characterized as δ-Ni_2Si intermetallic compounds by an optical diffractogram. As will be discussed in subsequent section, the dark phases in Fig. 3(a) were DPs with the fibre-shaped Ni_2Si precipitates which have also been called "cellular precipitates".

Fig. 4(a) shows TEM images on the microstructure around GB areas in the CP specimen. A band-like bright zone and a particle with a few tens of nanometres were observed along the GBs (refer to magnified views: Fig. 4(b) and (d)). Fig. 4(c) and (e) indicated that the bright zone and particle was the precipitate-free zone (PFZ) and NiSi intrametric compounds (heterogeneous precipitates), respectively. Along with NiSi, Ni_2Si intermetallics with a few tens of nanometres in size was also observed at GB areas [14]. The generation and subsequent growth of the heterogeneous precipitates (NiSi, Ni_2Si) at GBs were brought by the fast diffusion along the GBs. Accordingly, solute atoms' absorption near the heterogeneous precipitates should lead to the formation of PFZ.

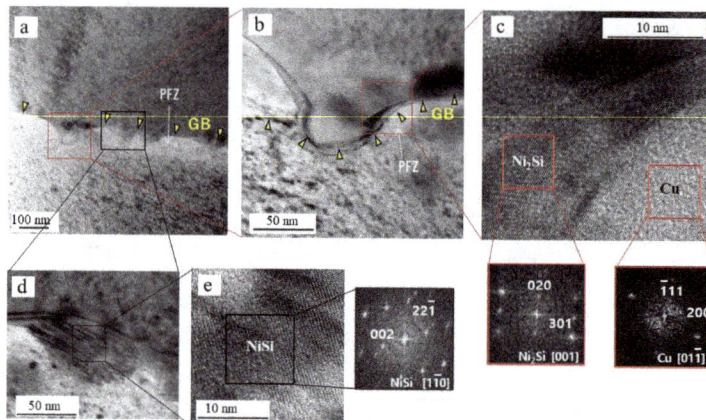

Figure 4:   TEM micrographs of GB areas in the CP specimen.

3.2 Microstructure of the DP specimen

Fig. 5(a) shows the microstructure of the DP specimen, which consists of almost all DP phases. Inclusions with a few micrometres in size were formed and most of which were along GBs. The colonies of fibre-shaped δ-Ni$_2$Si intermetallic compounds were observed inside the grain [15]. Yellow arrow heads in Fig. 5(b) indicate the DP–DP–phase boundaries in the grain. It has been established that DP is formed by the decomposition of supersaturated solid solution into solute-depleted matrix and precipitates across moving boundary [17]. Previous study [15] on the formation of DP structure has shown that after DP generation at the GBs, the DPs grew towered the grain interior with moving boundaries. Colonies of Ni$_2$Si fibre (Fig. 5(c)) with an extremely high aspect ratio were then formed within intergranular DPs. The DPs grew along a specific orientation in the Cu matrix, even though their growth orientations were not always uniform inside the same grain (parent grain). Although the formation of DPs is detrimental to the hardness of Cu–Ni–Si alloys, the electrical conductivity of DPs is superior to that of the normal aged alloy with the highest hardness (Fig. 2). The further reduction of Ni and Si solute elements in the Cu matrix during the DP formation induced the high electrical conductivity of DP specimen.

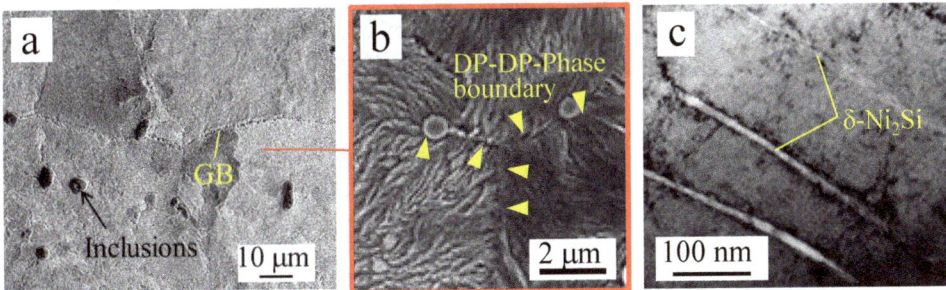

Figure 5:   Microstructure of the DP specimen. (a) OM micrograph; (b) DP–DP–phase boundaries characterized by colonies of fibre-shaped Ni$_2$Si with a specific orientation; and (c) TEM image of Ni$_2$Si fibre in the Cu matrix.

3.3 Microstructure of the cold-rolled DP specimen

Simultaneous increase in electrical conductivity and tensile strength of precipitate strengthened Cu–Ni–Si alloys is often extremely difficult, because a decrease in electric conductivity is inevitably caused by the available strengthening mechanism. Using Cu–6Ni–1.4Si–0.1Ti alloys, Han et al. [8] fabricated alloys with fully DP phases through a prolonged aging (overaging) process. The product was then mechanically rolled with 90% reduction in area. Such a rolled DP phases showed a simultaneous increase in electrical conductivity and tensile strength by 1.3 times compared with conventional precipitate-strengthened Cu alloys. These recent results suggest that Cu alloys with fully DP phases processed by severe plastic deformation have a great potential as electrical materials with high electrical conductivity and super-high strength. Here, the microstructure of DP phases subjected to cold-rolling was investigated. Accordingly, some of the specimens with fully DP phases were then cold-rolled down to 80% reduction in area (DPR specimen).

Fig. 6(a) shows the OM micrograph observed on the surface of DP specimen. The DP phases were composed of fibre-shaped Ni$_2$Si intrametric compound and Cu matrix (Fig. 6(c)).

Fig. 6(b) is the OM micrograph of the DPR specimen, showing stream lines along rolling direction. These lines were the array of Ni_2Si intermetallic fibres which tend to align along the direction of cold-rolling (Fig. 6(d)). The analysis [18] on the nanofiber in DP structure showed that the diameter of intermetallic fibre reduced from 14 to 7 nm as the result of 80% cold-rolling, while the average spacing between them reduced from 87 to 41 nm.

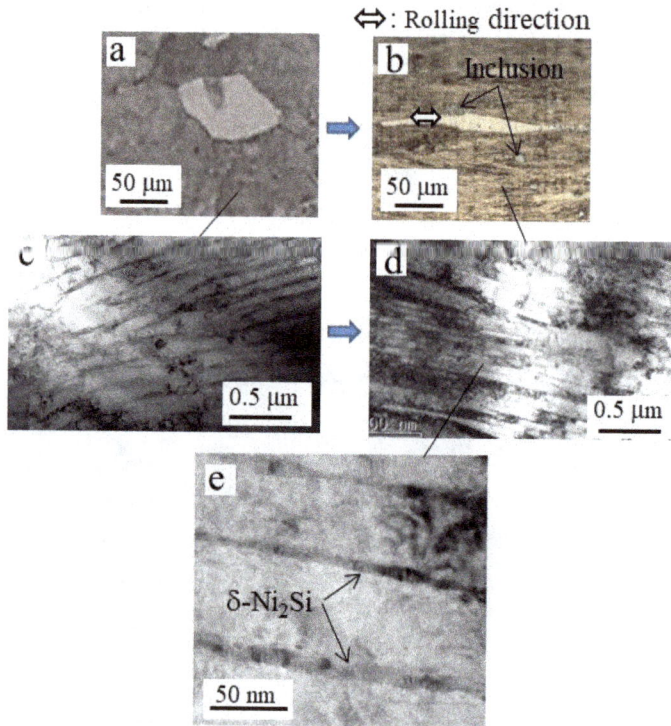

Figure 6: OM and TEM micrographs of the DPR specimen. (a) and (c) Microstructure before cold-rolling; (b) and (d) Microstructure after cold-rolling; and (e) Magnified view of Ni_2Si fibre after cold-rolling.

3.4 Tensile properties of precipitate strengthened Cu–6Ni–1.5Si alloy

The stress–strain curves of the CP, DP and DPR specimens are shown in Fig. 7. Indeed, the tensile strength of the DP specimen was lower than that of the CP specimen, however, the DPR specimen shows the highest tensile strength among the specimens tested. The highest tensile strength of DPR specimen was mainly caused by the decrease in the radii of aligned Ni_2Si nanofibers and the interdistance between them. It has been shown that an increase in tensile strength induced by work hardening is less than 60% of that brought by the nanofiber array formation in Cu–6Ni–1.4Si–0.1Ti alloys subjected to cold-rolling with 90% reduction in area [8]. The electrical conductivity of the CP and DP specimen was 25 and 44%IACS, respectively (Fig. 2), while that of the DPR specimen was 41%IACS. Consequently, the DPR specimen retained a superior combination between the tensile strength and electrical conductivity, while the tensile elongation was significantly reduced.

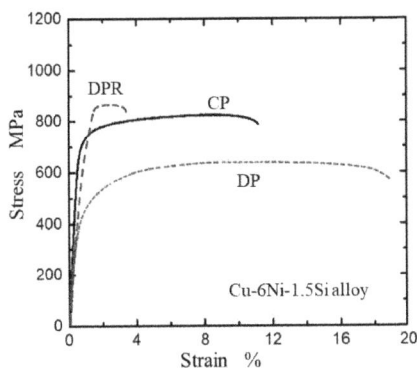

Figure 7: Stress–strain curves of the CP, DP and DPR specimens.

3.5 Fatigue characteristics of precipitate strengthened Cu–6Ni–1.5Si alloy

Previous studies [15], [18] indicated that the S–N plots of the CP specimen were comparable to that of the DP specimen despite a higher tensile strength for the CP specimen than that for the DP specimen, and the DPR specimen showed the highest fatigue strength among specimens tested. The fatigue limit stress at $N = 10^7$ cycles, σ_w, of the DPR specimen (σ_w= 300 MPa) was about 1.5 times of the CP (σ_w= 200 MPa) and DP (σ_w= 195 MPa) specimens. The ratios of σ_w, to the tensile strength, σ_u, was 0.35 and 0.34 for the DP and DPR specimen, respectively, while it was 0.25 for the CP specimen.

Fig. 8(a)–(c) shows the surface morphology around an initial crack formed by cyclic stressing at $\sigma_a = 400$ MPa. Fatigue crack in the CP specimen was initiated at GBs, followed by propagation along a favourable crystallographic slip orientation of the grain not along GBs. On the DP specimen, crack was initiated along GBs. Although the crack path just after the initiation was along GBs, the crack passed through both GBs and DP phases with further stressing. The number of cycles to create a grain-sized crack ($l \approx 100$ µm) was 3×10^4 for the CP specimen and 7×10^4 for the DP specimen, respectively. It has been shown [14] that GB areas of the CP specimen are weakened because of PFZ formation (Fig. 4(a)). An easy GB-crack initiation of the CP specimen was attributed to weakened GB areas despite the high tensile strength compared with the DP specimen. A 100-µm long crack in the DPR specimen was initiated after 1×10^5 repetitions that was significantly larger than that for the CP and DP specimens. The arrays of finely distributed Ni_2Si nanofibers brought the high strength of DPR specimen and the interface of nanofibers possessed a coherency with Cu matrix [7], resulting in a drastic increase in crack initiation resistance.

Fig. 9(a) shows the crack growth curve (ln$l$ vs. N) of the CP and DP specimens at $\sigma_a = 300$ MPa. The initiation life of grain-sized crack in the CP specimen was shorter than the DP specimen (Fig. 8(a) and (b)). On the contrary, the crack growth rate was slower in the CP than the DP specimen. Accordingly, comparable S–N plots between the CP and DP specimens were attributed to the shorter life on grain-sized crack generation and longer life on crack growth of the CP specimen. Fig. 9(b) shows the comparison of crack growth curves between the DP and DPR specimens at $\sigma_a = 400$ MPa. The slop of growth curve for the DP specimen was steeper than that for the DPR specimen, indicating an enhanced crack growth resistance in the DPR specimen.

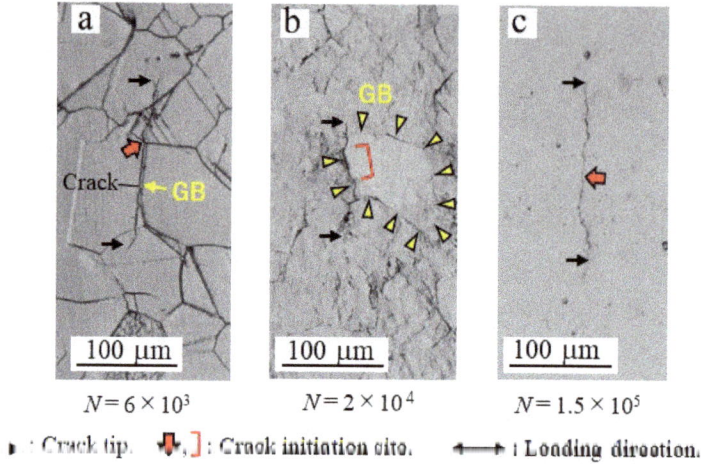

Figure 8: The surface damage around crack initiation site at $\sigma_a = 400$ MPa. (a) The CP specimen; (b) The DP specimen; and (c) The CPR specimen.

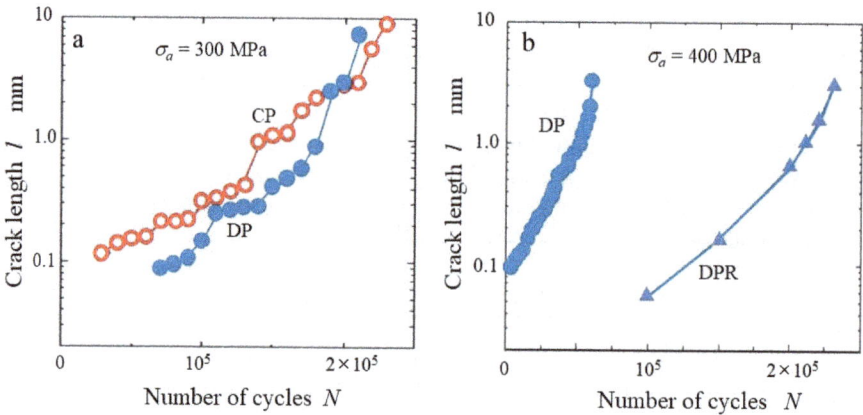

Figure 9: Crack growth curves ($\ln l$ vs. N correlation). (a) The CP and DP specimen at $\sigma_a = 300$ MPa; and (b) The DP and DPR specimens at $\sigma_a = 400$ MPa.

Fig. 10(a) shows the dependency of crack growth rate (CGR, dl/dN) on crack length l for the DPR specimen. Here, the CGR was calculated from the growth curves represented by smoothed curves which pass through the average of each set of plots. For every constant stress range, a straight line could be drawn approximately for crack length smaller than 2 mm, while the slop of the lines depended on the stress amplitude. For the range of $\sigma_a \geqq 400$ MPa, the slope is about 1, indicating that the dl/dN is proportional to l. However, the slope of line at $\sigma_a = 350$ MPa was steeper than that for high stress ranges ($\sigma_a \geqq 400$ MPa) and was nearly equal to 2. Extended study on the dl/dN vs. l correlation of the CP and DP specimens showed that the dl/dN was nearly proportional to l at the stress range of $\sigma_a \geqq 300$ MPa.

It has been established that stress intensity factor range (SIFR, ΔK) is applicable to estimate the growth rate of fatigue crack propagating under the condition of small-scale yielding (a long/large crack at a comparatively low stress). The SIFR cannot estimate the growth rate of a small/short crack (e.g., $l < 1$ mm) [19], [20], unlike a large crack growing under a low cyclic stress. Nisitani et al. [21] and Goto and Nisitani [22] proposed eqn (1) for determining the growth rate of a mechanically small crack propagating under the large-scale yielding condition:

$$\frac{dl}{dN} = C_1 \sigma_a^n l \tag{1}$$

where C_1 and n are material constants. The studies on small crack growth behaviour in various metals and alloys, such as carbon steels [21], low alloy steels [22], aluminium alloys [23], Ni-based superalloys [24], copper [25] and copper alloys [26], indicated that dl/dN of a small surface-crack in smooth specimens is determined by the term $\sigma_a^n l$.

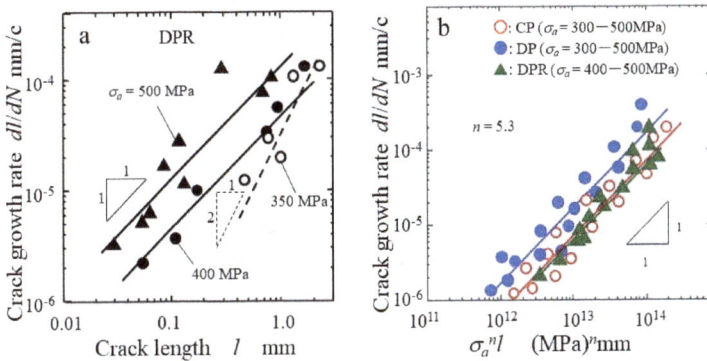

Figure 10: The crack growth data. (a) The dl/dN vs. l relationship of the DPR specimen; and (b) The dl/dN vs. $\sigma_a^n l$ relationship of the CP, DP and DPR specimens.

Fig. 10(b) shows the growth data of a small surface-crack with the dl/dN vs. $\sigma_a^n l$ correlation. The value of n in eqn (1) was 5.3 in all specimens. Excepting a lower stress amplitude ($\sigma_a < 300$ MPa for the CP and DP specimens, and $\sigma_a < 400$ MPa for the DPR specimen), the CGR could be uniquely determined by $\sigma_a^n l$, supporting the validity of eqn (1) for estimation of growth rate of a small crack. The crack propagation resistance of the DPR specimen was nearly equal to that of the CP specimen. Therefore, a longer fatigue life of the DPR specimen was mainly caused by an enhanced crack initiation resistance rather than crack propagation. At a low stress amplitude, the SIFR should be applicable. The slope of the dl/dN vs. l correlation at $\sigma_a = 350$ MPa in the DPR specimen was nearly equal to 2 (Fig. 10(a)), suggesting the applicability of Paris law: $dl/dN = C\Delta K^4$. Bellinia et al. [27] conducted fatigue tests of Cu–Ni–Si alloy using the CT specimen. They showed the lower ΔK_{th} and a high CGR for a low ΔK range, followed by a well-developed Paris stage in a range of $\Delta K > 6$ MPa·m$^{0.5}$.

The fracture surfaces (1.0 mm beneath the specimen surface) of the CP, DP and DPR specimens are shown in Fig. 11. The fracture surface of the CP specimen (Fig. 11(a)) had the rough surface characterized by intergranular and transgranular facets. Such rough surface was formed as the result of crack propagation along both favourable slip planes of grains and

GBs. The transgranular fracture surface with the presence of some areas characterized by intergranular facets has been observed in Cu–Ni–Si alloy with a fine and dispersed Ni_2Si phase [27]. The fracture surface of the DP specimen was comparatively flat (Fig. 11(b)) which contrasted well with the rough fracture surface of the CP specimen. The flat surface appeared to be caused by a variety of Ni_2Si fibre orientation of DP colonies formed inside the parent grain, unlike the CP specimen with a single crystallographic slip orientation in the grain. The fracture surface of the DPR specimen is shown in Fig. 11(c). The magnified view of a highlighted area in Fig. 11(c) is likely to assist the crack growth by forming striations.

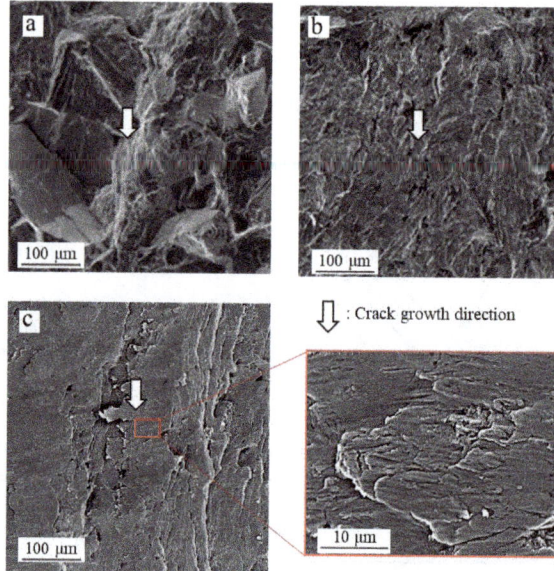

Figure 11: The fracture surface at 1.0 mm beneath the surface. (a) The CP specimen (σ_a = 300 MPa); (b) The DP specimen (σ_a = 300 Mpa); and (c) The DPR specimen (σ_a = 350 MPa).

4 CONCLUSIONS

The main findings of this study of the microstructure and fatigue properties on precipitate strengthened Cu–6Ni–1.5Si alloys can be summarized as follows.

1. The normal aged specimen (CP) had the matrix strengthened by disk-shaped δ-Ni_2Si intermetallics with nano-size diameter. The microstructure of overaged specimen (DP) showed a cellular component composed of fibre-shaped δ-Ni_2Si in the Cu matrix. The fibre-shaped δ-Ni_2Si intermetallics were elongated in the DP specimen subjected to the cold-rolling (80%) (DPR). The diameter of δ-Ni_2Si fibre decreased from 14 to 7 nm, while the average spacing between them decreased from 87 to 41 nm.
2. The DRP specimen showed the highest tensile strength among the specimens tested: 1.5 and 1.06 times of the DP specimen and CP specimen, respectively.
3. The fatigue limit stress at 10^7 cycles of the DPR specimen was 1.5 times of the CP and DP specimens. The initiation resistance of a grain-sized crack ($l \approx 100$ μm) in the DPR specimen was drastically enhanced.

4. At the stress range of $\sigma_a \geqq 300$ MPa for CP/DP and $\sigma_a \geqq 400$ MPa for DPR specimen, the growth rate of a small crack can be determined by a term $\sigma_a^n l$. The value of n was 5.3 regardless of a difference in the specimens. The crack growth resistance of the DPR specimen was nearly equal to that of the CP specimen.

5. The increased tensile and fatigue strengths of the DPR specimen were attributed to a reduced spacing between $\delta\text{-Ni}_2\text{Si}$ fibres along with work hardening. Consequently, the DPR specimen was superior to the CP specimen in all of electrical conductivity, tensile strength and fatigue resistance.

ACKNOWLEDGEMENTS

This study was supported by a Grant-in-Aid for Scientific Research (Kiban-B) (KAKENHI: No. 18H01340) and for Encouragement of Scientists (No. 18H00244) from the Japan Society for the Promotion of Science, as well as the National Research Foundation of Korea (NRF) grant funded by the Korea government (MSIP) 2020M3D1A2098962. The authors are very grateful to the members of the Strength of Materials Laboratory of Oita University for their excellent experimental assistance. Thanks are also extended to the members of the Korea Institute of Materials Science, for the fabrication of the Cu–Ni–Si system alloys.

REFERENCES
[1]   Corson, M.G., Electrical conductor alloys. *Electr. World*, **89**, pp. 137–139, 1927.
[2]   Locker, S.A. & Noble, F.W., Precipitate microstructure in a Cu-Ni-Si alloy. *J. Mater. Sci.*, **29**, pp. 218–226, 1994.
[3]   Fujiwara, H., Sato, T. & Kamio, A., Effect of alloy composition on precipitation behavior in Cu-Ni-Si alloys. *J. Jpn. Inst. Metals.*, **62**, pp. 301–309, 1998.
[4]   Hu, T., Chen, J.H., Liu, J.Z., Liu, Z.R. & Wu, C.L., The crystallographic and morphological evolution of the strengthening precipitates in Cu–Ni–Si alloys. *Acta Mater.*, **61**, pp. 1210–1219, 2013.
[5]   Srivastava, V.C., Schneider, A., Ojha, V., Uhlenwinkel, S.N. & Bauckhage, K., Age-hardening characteristics of Cu–2.4 Ni–0.6 Si alloy produced by the spray forming process. *J. Mater. Process. Technol.*, **147**, pp. 174–180, 2004.
[6]   Monzen, R. & Watanabe, R., Microstructure and mechanical properties of Cu–Ni–Si alloys. *Mater. Sci. Eng. A*, **483–484**, pp. 117–119, 2008.
[7]   Favez, D., Wagnière, J.D. & Rappaz, M., Au–Fe alloy solidification and solid-state transformations. *Acta Mater.*, **58**, pp. 1016–1025, 2010.
[8]   Han, S.Z. et al., Increasing strength and conductivity of Cu alloy through abnormal plastic deformation of an intermetallic compound. *Sci. Rep.*, **6**, 30907, 2016.
[9]   Semboshi, S., Sato, S., Iwase, A. & Takasugi, T., Discontinuous precipitates in age-hardening Cu–Ni–Si alloy. *Mater. Charact.*, **115**, pp. 39–45, 2016.
[10] Lockyer, S.A. & Noble, F.W., Fatigue of precipitate strengthened Cu–Ni–Si alloy, *Mater. Sci. Technol.*, **15**, pp. 1147–1153, 1999.
[11] Sun, Z., Laitem, C. & Vincent, A., Dynamic embrittlement during fatigue of a Cu–Ni–Si alloy. *Mater. Sci. Eng. A*, **528**, pp. 6334–6337, 2011.
[12] Fujii, T., Kamio, H., Sugisawa, Y., Onaka, S. & Kato, M., Cyclic softening of Cu–Ni–Si alloy single crystals under low-cycle fatigue. *Mater. Sci. Forum*, **654–656**, pp. 1287–1290, 2010.
[13] Delbove, M., Vogt, J.-B., Bouquere, J., Soreau, T. & Primaux, F., Low cycle fatigue behavior of a precipitation hardened Cu–Ni–Si alloy. *Int. J. Fatigue*, **92**, pp. 313–320, 2016.

[14] Goto, M., Han, S.Z., Lim, S.H., Kitamura, J., Fujimura, T., Ahn, J.-H., Yamamoto, T., Kim, S. & Lee, J., Role of microstructure on initiation and propagation of fatigue cracks in precipitate strengthened Cu–Ni–Si alloy. *Int. J. Fatigue*, **87**, pp. 15–21, 2016.

[15] Goto, M. et al., Microstructure-dependent fatigue behavior of aged Cu-6Ni-1.5Si alloy with discontinuous/cellular precipitates. *Mater. Sci. Eng. A*, **747**, pp. 63–72, 2019.

[16] Gholami, M. et al., Influence of grain size and precipitation hardening on high cycle fatigue performance of CuNiSi alloys. *Mater. Sci. Eng. A*, **684**, pp. 524–533, 2017.

[17] Findik, F., Discontinuous (cellular) precipitation. *J. Mater. Sci. Lett.*, **17**, pp. 79–83, 1998.

[18] Goto, M. et al., Simultaneous increase in electrical conductivity and fatigue strength of Cu-Ni-Si alloy by utilizing discontinuous precipitates. *Mater. Lett.*, **288**, 129353, 2021.

[19] Miller, K.J. & de los Rios, E.G. (eds), *The Behaviour of Short Fatigue Cracks*, EGF-Pub 1, MEP, 1986.

[20] Ravichandran, K.S., Ritchie, R.O. & Murakami, Y. (eds), *Small Fatigue Cracks. Mechanics, Mechanisms and Applications*, Elsevier, 1999.

[21] Nisitani, H., Goto, M. & Kawagoishi, N., A small-crack growth law and its related phenomena. *Eng. Fract. Mech.*, **41**, pp. 499–513, 1992.

[22] Goto, M. & Nisitani, H., Fatigue life prediction of heat-treated carbon steels and low alloy steels based on a small-crack growth law. *Fatigue Fract. Eng. Mater. Struct.*, **17**, pp. 171–185, 1994.

[23] Goto, M. & Kawagoishi, N., Growth behavior and crack distribution characteristics of small surface cracks of age-hardened Al-alloy 2017-T4. *J. Society Mater. Sci. Jpn.*, **45**, pp. 675–679, 1996.

[24] Goto, M. & Knowles, D.M., Initiation and propagation behavior of microcracks in Ni-based superalloy Udimet 720 LI. *Eng. Fract. Mech.*, **60**, pp. 1–18, 1998.

[25] Goto, M. et al., High-cycle fatigue strength and small-crack growth behavior of ultrafine-grained copper with post-ECAP annealing. *Eng. Fract. Mech.*, **110**, pp. 218–232, 2013.

[26] Goto, M. et al., Behaviour of small fatigue cracks in Cu-5.5Ni-1.28Si alloy round-bar specimens. *WIT Trans. Eng. Sci.*, **125**, pp. 3–13, 2019.

[27] Bellinia, C., Brotzu, A., Di Cocco, V., Felli, F., Iacoviello, F., Pilone, D., Fatigue crack propagation mechanisms in C70250 and CuCrZr copper alloys. *Procedia Struct. Integrity*, **26**, pp. 330–335, 2020.

DESIGN FOR ADDITIVE MANUFACTURING: IS IT AN EFFECTIVE ALTERNATIVE? PART 1 – MATERIAL CHARACTERIZATION AND GEOMETRICAL OPTIMIZATION

FRANCO CONCLI[1], MARGHERITA MOLINARO[2] & ELEONORA RAMPAZZO[1]
[1]Faculty of Science and Technology, Free University of Bolzano/Bozen, Italy
[2]Polytechnic Department of Engineering and Architecture, University of Udine, Italy

ABSTRACT

Additive manufacturing (AM) is becoming a more and more widespread (and trendy) approach. Its flexibility and capability to manufacture any topology has opened new possibilities: AM could lead to significant performance improvements thanks to the exploitation of lattice or reticular structures as partial replacement of the traditional solid design. The potential of this technology knows no bounds. However, in the real world, the lower performances of the materials and the high manufacturing costs significantly restrict the fields of application for which the adoption of AM results effective. In this context, the mechanical static and fatigue properties of a 17-4 PH stainless steel produced via AM were experimentally measured and compared with those of the wrought material to quantify the performance reduction. Based on these data, three components, namely a hip prosthesis, a blow plastic bottle die, and an automotive gear were selected as representative examples to show the pros and contra of AM. The three components were chosen because they belong to three quite dissimilar fields and are produced in different batch sizes. The three original designs were specifically optimized for AM by means of finite element (FE) simulations. The new solutions fulfil the strength requirements of the original parts showing at the same time reduced weights and inertias. The traditional and new designs were compared in terms of production times and costs to quantify the real benefits of AM for different applications.
Keywords: additive manufacturing, FEM, optimization.

1 INTRODUCTION

The main goal of this work is to investigate the effectiveness of additive manufacturing (AM) as alternative to traditional machining. To have a more complete overview, a first part of the research was focused on the characterization of an AM 17-4 PH stainless steel and the comparison of its mechanical properties with those of the wrought counterpart.

In this regard, two series of samples were manufactured via laser powder bed fusion (L-PBF ISO/ASTM 52900) according to ASTM-E606 [1].

Some authors have proven that the mechanical properties of this steel manufactured via L-PBF are comparable or slightly above those of wrought counterpart [2]. This is mainly due to the rapid cooling rates that characterize the additive manufactured parts, leading to a finer micro-structure [3]. The high solidification speeds characteristic to the L-PBF process, in fact, impede the formation of the martensite phase in the as-built material leading to a metastable austenitic micro-structure [4]–[6]. Moreover, typical porosities which comes from the production process can significantly impact the mechanical properties of the material [7]–[10]. The voids, in fact, act as crack nucleation sites. While the characteristic porosities of AM parts could negatively affect the mechanical properties, AM shows the interesting capability to produce any kind of geometry including lattice structures and reticula, removing the geometrical constrains of the traditional manufacturing. This incredible flexibility raises another important point: lattice and reticular structures rely on thin struts that, under operation, could locally exceed the yielding leading to local plastic deformations [11]–[14].

WIT Transactions on Engineering Sciences, Vol 130, © 2021 WIT Press
www.witpress.com, ISSN 1743-3533 (on-line)
doi:10.2495/CMEM210081

To be able to consider this effect during the optimization of the structures (Design for AM), the knowledge of the cyclic behavior is fundamental. In order to have reliable data, low-cycle-fatigue (LCF) tests were performed leading to the fine-tuning of the Ramberg–Osgood (RO) model. Moreover, the tests lead to the calibration of the Basquin–Manson–Coffin (BMC) curve [15].

While the high-cycle-fatigue (HCF) failure mechanism is promoted by the cyclic elastic loading and governed by parameters such the stress concentration factor, the surface roughness and the mean stress level, in the LCF range, the failure could be related to the plastic deformations which are responsible for the trans-granular cracks initiation and propagation [16], [17].

The testing campaign relies on two series of samples. The first series was tested in the as-build condition, while the second one was further machined to achieve a better surface finishing. This knowledge is fundamental considering that in most applications the AM lattice/reticular structures are set into operation in the as-built condition – finishing operations are not possible.

Once the material characterization was completed, the successive step of this research was focused on the geometry optimization, namely the design for AM. The optimization was performed numerically via finite element analysis (FEA). The variables for the optimization were the struts diameter and cell size, while the optimization objectives were the minimization of the mass and the deformation. The optimization was applied to three completely different components, namely a hip prosthesis, a blowing bottle mold and an automotive gear. These three components were chosen because they belong to three quite dissimilar fields and are produced in different batch sizes (these aspects will be fundamental for the economical evaluations). The three optimized solutions fulfil the strength requirements of the original parts showing at the same time reduced weights and inertias. Finally, the traditional and new designs were compared in terms of production times and costs to quantify the real benefits of introducing AM.

2 MATERIAL CHARACTERIZATION

Fig. 1 shows the samples used for the quasi-static (QS) tests and for the LCF tests. The geometry for the QS test consists of a smooth cylinder whose active part has a length of 35 mm and a nominal diameter of 5 mm. For what concerns the LCF samples, both the as-built and the wrought material series have a final geometry consisting in an active length of 4.28 mm and a nominal diameter of 2 mm. The geometry was designed according to the ASTM E606 standard [1].

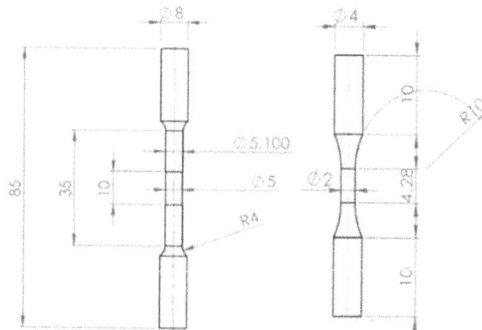

Figure 1: Specimens used for QS and LCF tests.

Table 1 shows the chemical composition of the metal powder as provided by the producer.

Table 1: Chemical composition of powder 17-4 PH SS.

C	Si	Mn	P	S	Cr	Ni	Mo
0.036	0.78	0.33	0.009	0.004	16.2	4.02	0.002

2.1 Quasi static tests

The QS tests were performed on a STEPLab UD04 testing machine (Fig. 2) owned by the Free University of Bolzano/Bozen. The machine can apply forces up to 5 kN.

Figure 2: Tensile machine.

During the tests, the crosshead displacement was equal to 0.1 mm/min. Tests were performed at room temperature. The typical cone-cup shaped ductile fracture was observed on the specimens. The AM material shows a Yielding of 735 MPa, slightly below the one of the wrought 17-4 PH Stainless Steel (980 MPa) [18]–[21].

2.2 Low-cycle-fatigue tests

Strain-controlled fatigue tests were performed on the same testing machine used for the QS tests with a strain ratio $R_\varepsilon = -1$. The testing frequency was set to 0.1 Hz. This value, the minimum value prescribed by the ASTM E606 [1], was kept constant during the tests. Higher speeds may affect the results promoting an increment of the temperature of the sample and a modification of the mechanical properties.

The stabilized cycles at the different stain levels were interpolated (Fig. 3) by means of the Ramberg–Osgood (RO) equation

$$\varepsilon_a = \varepsilon_{ae} + \varepsilon_{ap} = \frac{\sigma_a}{E} + \left(\frac{\sigma_a}{K'}\right)^{\frac{1}{n'}} -, \tag{1}$$

where ε_{ae} and ε_{ap} are the elastic- and plastic-strain amplitudes, σ_a the stress amplitude, E the elastic module, and K' and n' are constants that depend on the material [22]. In this form K' and n' are not the same as the constants commonly seen in the Hollomon equation [23].

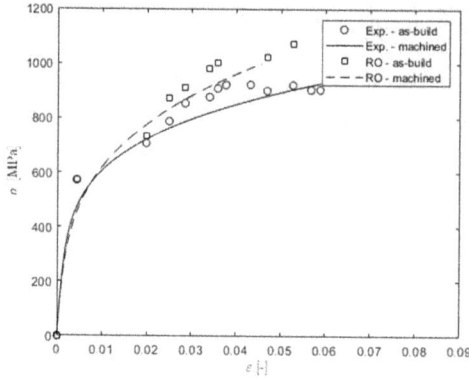

Figure 3: Ramberg–Osgood σ-ε as-built/machined curves comparison [24], [25]

The parameters of the RO equations are reported in Table 2.

Table 2: Parameters of the Ramberg–Osgood equation.

Sample	K' [MPa]	n' [−]	E [MPa]
As-built	1,705	0.2092	210,000
Machined	2,392	0.2730	210,000

Even after the stabilization, tests were continued up to rupture of the sample to obtain data to tune the Basquin–Coffin–Manson (BCM) fatigue equation.

$$\varepsilon_a = \frac{\Delta\varepsilon}{2} = \frac{\Delta\sigma'_f}{E}(2N)^b + \varepsilon'_f(2N)^c . \tag{2}$$

The BCM equation can be split in two parts: the first refers to Basquin's formulation and describes the elastic share of the deformation, while the second term, the so-called Manson–Coffins' one, describes the plastic part. The first two parameters σ'_f and b are the fatigue strength coefficient and the fatigue strength exponent respectively, while ε'_f and c are the fatigue ductility coefficient and the fatigue ductility exponent. To calibrate these values, the ASTM 739 standard was used [26].

The calculation procedure prescribes to determine the Basquin coefficients (σ'_f and b) starting from the yield stress σ_Y and the high-cycle-fatigue limit σ_F [27] with a linear interpolation. The Basquin equation can be used to separate the plastic deformation ε_{ae} from the total one.

$$\varepsilon_{ap} = \varepsilon_a - \varepsilon_{ae} = \varepsilon'_f(2N)^c . \tag{3}$$

To calculate ε'_f and c, eqn (3) can be rewritten as

$$\log(N) = \hat{A} + \hat{B}\log(\varepsilon_{ap}) . \tag{4}$$

The maximum likelihood estimators of A and B are following

$$\hat{A} = \overline{Y} - \hat{B}\overline{X}, \tag{5}$$

$$\hat{B} = \frac{\sum (x_i - \overline{X})(y_i - \overline{Y})}{\sum (x_i - \overline{X})^2}, \tag{6}$$

where $\overline{X}$ and $\overline{Y}$ are the averaged values of the dependent variable x_i ($\log \varepsilon_{ap}$) and independent variable y_i ($\log N$). Eqn (4) can be transformed to:

$$\varepsilon_{ap} = 10^{-\frac{\hat{A}}{\hat{B}}} \left(\frac{1}{2}\right)^{\frac{1}{\hat{B}}} (2N)^{\frac{1}{\hat{B}}} \tag{7}$$

Therefore, the maximum likelihood estimators of the Coffin–Manson coefficients ε'_f and c (eqn (3)) can be calculated as

$$\varepsilon'_f = 10^{-\frac{\hat{A}}{\hat{B}}} \left(\frac{1}{2}\right)^{\frac{1}{\hat{B}}} \text{ and } c = 1/\hat{B}. \tag{8}$$

The values of the calibrated BCM models are reported in Table 3.

Table 3: Parameters of the Basquin–Coffin–Manson equation.

Sample	b	c	σ'_f	ε'_f
As-built	−0.0250	−0.2365	390	0.1263
Machined	−0.0160	−0.2204	426	0.1299

Fig. 4 shows how the machined samples better perform with respect to the as-build ones.

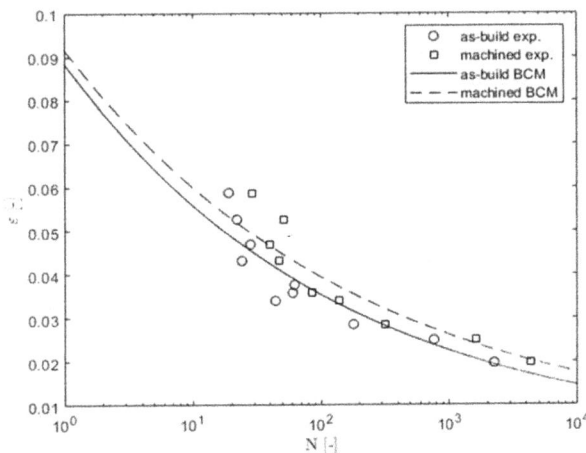

Figure 4: Resulting BMC curves.

3 DESIGN FOR AM

The flexibility of AM [28] opens new scenarios in terms of optimization and re-design of existing components. In the next sections, three examples are studied. The external shape of the parts as well as the loads were considered as constraints. The optimization relies on the possibility offered by AM to substitute the solid design with reticular/lattice structures. In this way the weight and the inertias of the components could be potentially reduced with advantages also in terms of material saving and environmental impact. However, while with AM every design is feasible from a technological point of view, only through a specific optimization it is possible to maximize the benefits.

The approach used in this project foresees to optimize an elementary cell to achieve the requirements in terms of maximum stresses and deformations required by the three considered applications. After a preliminary screening based on literature, the BCC, the BCCZ and the FCC cell topologies (Fig. 5) were selected as the most appropriate for the successive steps. A parametric model of the cell was created in the open-source environment Salome-Meca/Code_Aster [29]. This Finite Element (FE) solver [30] was coupled with Dakota (Design Analysis Kit for Optimization and Terascale Applications) [31]. The mesh element size was selected after a mesh sensitivity analysis resulting in 0.15 mm.

BCC   BCCZ   FCC

Figure 5: Cell topologies.

The optimization procedure works as follows. Starting from the unitary loads characteristics of the 3 studied systems – hip prosthesis, blowing mold and automotive gear, a first simulation was performed (for each system and cell topology combination). Based on the results in terms of maximum equivalent stress in the reticula and maximum deformation, the struts diameters and the elementary cell size were automatically modified, and the next optimization step performed. Two objective functions (mass minimization and stiffness maximization) were used. The new parameters for the successive simulations were selected according to the MOGA (Multi-objective Genetic Algorithm), which is a global optimization method that does Pareto optimization for multiple objectives. It supports general constraints and a mixture of real and discrete variables.

This approach was applied to the three different system considered. While the hip prosthesis was chosen as representative of a one-sample batch and the die mold of a small batch, the gear represents the typical series production.

3.1 Hip prosthesis

Since we are talking about movement, gravity and weights, the loads are dynamic and difficult to predict. At least for the standard walking, the resultant force along the entire footstep cycle lays in a direction that remains inside a small cone of action, in the superior zone of the acetabulum (Fig. 6).

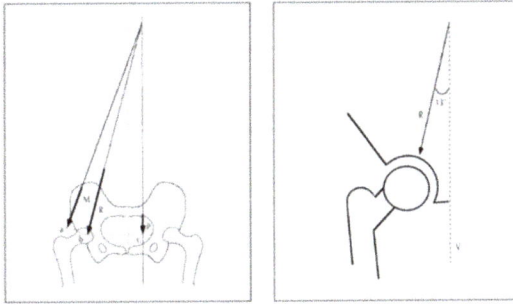

Figure 6: Loads and their angle of action on the acetabulum.

In case of monopodial loading, the force results 7–10 times the body weight of a person. Considering an average person weight of 85 kg, the maximum load could be quantified in 6,670 N.

The area of contact between the femur head and the acetabulum cavity is around 3,000 mm² Therefore, the contact pressure that should be used as input for the optimization is 2.22 MPa.

3.1 Bottle blowing mold

The second considered object is a mold used to produce plastic bottles (Fig. 7).

The material of the bottles is Polyethylene terephthalate, better known as PET, and it is obtained from the reaction of two chemicals known as ethylene glycol (EG) and purified terephthalic acid (PAT). The production process consists of blowing pressured air into the preform letting the sides stretch until they stick to the die. The pressure can be estimated in 3.6 MPa.

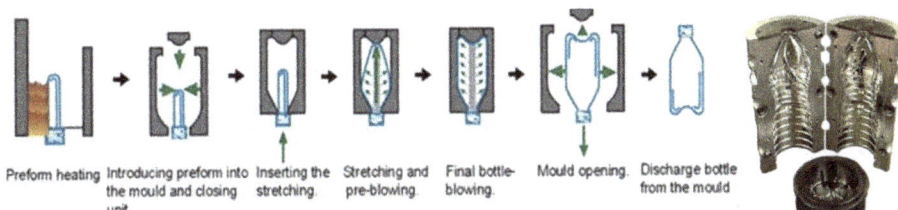

Figure 7: Stretch blow molding process step and mold geometry.

3.2 Automotive gear

The last mechanical component considered in this analysis is the fifth gear on the primary shaft of a Fiat Grande Punto 1200 cc gearbox. The choice was meant to include in the analysis a mass production component. For the maximum torque (377 Nm), the shear stress in the gear rim results equal to about 7 MPa.

All the above-mentioned levels of stress are significantly below the fatigue limit of any commercial steel and far below these of the 17-4 PH SS. This confirms that all the three considered components could be redesigned and lightened.

4 RESULTS OF THE OPTIMIZATION

4.1 Hip prosthesis

The optimization of the three cell topologies for the hip prosthesis leads to the results shown in Table 4.

Table 4: Optimized parameters for hip prosthesis.

Cell topology	Strut radius [mm]	Cell size [mm]
BCC	0.90	3.0
BCCZ	0.97	3.8
FCC	0.87	3.0

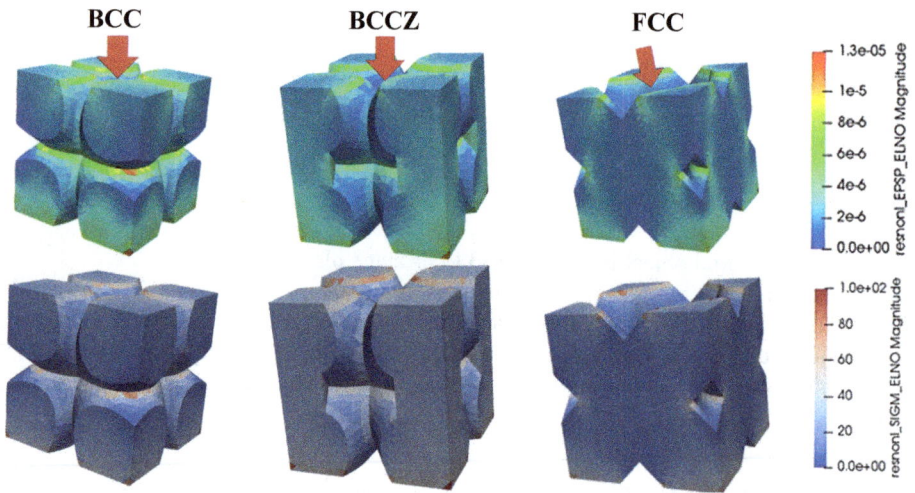

Figure 8: Optimized cells: strain and stress fields of hip prosthesis.

4.2 Bottle blowing mold

The optimization of the three cell topologies for the die mold leads to the results shown in Table 5.

Table 5: Optimized parameters for bottle blowing mold.

Cell topology	Strut radius [mm]	Cell size [mm]
BCC	0.99	3.3
BCCZ	0.88	3.0
FCC	0.87	3.0

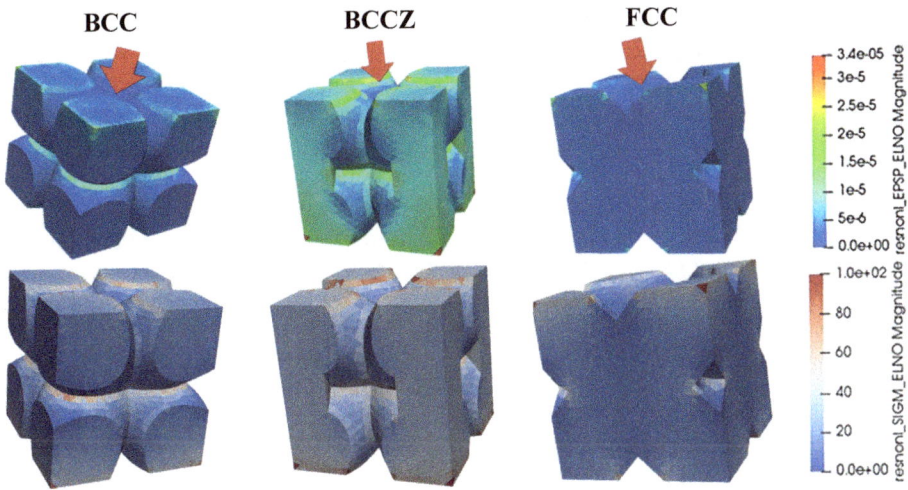

Figure 9: Optimized cells: strain and stress fields of bottle blowing mold.

4.3 Automotive gear

The optimization of the three cell topologies for the automotive gear leads to the results shown in Table 6.

Table 6: Optimized parameters for automotive gear.

Cell topology	Strut radius [mm]	Cell size [mm]
BCC	0.90	3.3
BCCZ	0.88	3.2
FCC	0.80	3.0

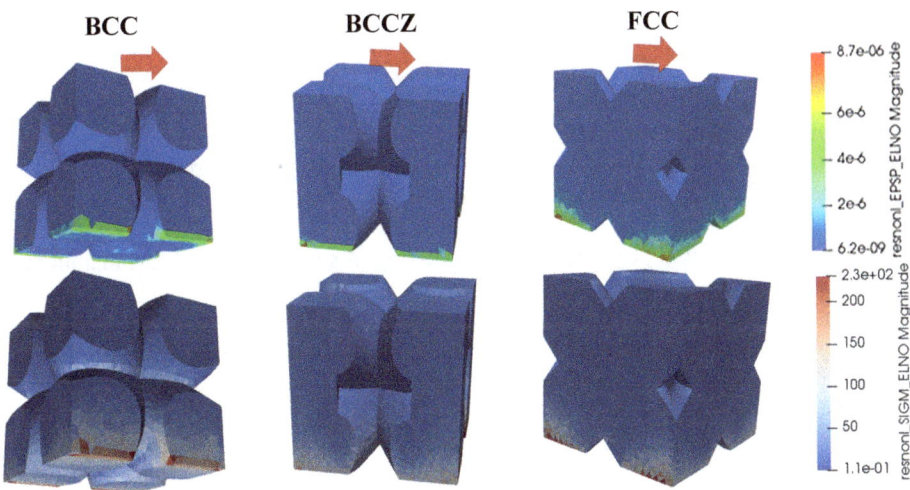

Figure 10: Optimized cells: strain and stress fields of automotive gear.

5 DISCUSSIONS AND CONCLUSIONS

The first step needed to perform this activity was to characterize the AM material to base the optimization on reliable data.

Three components belonging to three quite dissimilar fields and produced in different batch sizes were considered. Their solid design was replaced by optimized lattice structure, ensuring the same reliability and maximum deformation combined with a significantly reduced weight and inertia. Three reticular cell topologies were selected, BCC, BCCZ and FCC. Their strut diameters and cell size were optimized based on FEM simulations combined with a Multi-objective Genetic Algorithm. In the second part of this paper, based on the results of this study, an economical evaluation of the effectiveness and convenience of AM will be made.

REFERENCES

[1]   ASTM, Standard test method for strain-controlled fatigue testing. *E606/E606M-12*, vol. 96, no. 2004, pp. 1–16, 2004.
[2]   Cheruvathur, S., Lass, E.A. & Campbell, C.E., Additive manufacturing of 17-4 PH stainless steel: Post-processing heat treatment to achieve uniform reproducible microstructure. *JOM*, 2016.
[3]   S. Steel A. 17-4™ Precipitation ATI Technical Data Sheet and H. Alloy, Stainless Steel AL 17-4™ Precipitation Hardening Alloy. Allegheny Technologies, 2006.
[4]   Hsiao, C.N., Chiou, C.S. & Yang, J.R., Aging reactions in a 17-4 PH stainless steel. *Mater. Chem. Phys.*, **74**(2), pp. 134–142, 2002.
[5]   Rafi, H.K., Pal, D., Patil, N., Starr, T.L. & Stucker, B.E., Microstructure and mechanical behavior of 17-4 precipitation hardenable steel processed by selective laser melting. *J. Mater. Eng. Perform.*, **23**(12), pp. 4421–4428, 2014.
[6]   Viswanathan, U.K., Banerjee, S. & Krishnan, R., Effects of aging on the microstructure of 17-4 PH stainless steel. *Mater. Sci. Eng. C*, **104**, pp. 181–189, 1988.
[7]   Yadollahi, A., Shamsaei, N., Thompson, S.M., Elwany, A. & Bian, L., Mechanical and microstructural properties of selective laser melted 17-4 ph stainless steel. *ASME International Mechanical Engineering Congress and Exposition, Proceedings (IMECE)*, vol. 2A-2015, 2015.
[8]   Wu, J.H. & Lin, C.K., Influence of high temperature exposure on the mechanical behavior and microstructure of 17-4 PH stainless steel. *J. Mater. Sci.*, **38**(5), pp. 965–971, 2003.
[9]   Mirzadeh, H. & Najafizadeh, A., Aging kinetics of 17-4 PH stainless steel. *Mater. Chem. Phys.*, **116**(1), pp. 119–124, 2009.
[10] Luecke, W.E. & Slotwinski, J.A., Mechanical properties of austenitic stainless steel made by additive manufacturing. *J. Res. Natl. Inst. Stand. Technol.*, **119**, pp. 398–418, 2014.
[11] Lozanovski, B. et al., Computational modelling of strut defects in SLM manufactured lattice structures. *Mater. Des.*, **171**, 2019.
[12] Ren, X., Shen, J., Tran, P., Ngo, T.D. & Xie, Y.M., Design and characterisation of a tuneable 3D buckling-induced auxetic metamaterial. *Mater. Des.*, **139**, pp. 336–342, 2018.
[13] Ren, X., Shen, J., Tran, P., Ngo, T.D. & Xie, Y.M., Auxetic nail: Design and experimental study. *Compos. Struct.*, **184**(Oct. 2017), pp. 288–298, 2018.

[14] Yang, L., Harrysson, O., West, H. & Cormier, D., Correction to: Modeling of uniaxial compression in a 3D periodic re-entrant lattice structure. *J. Mater. Sci.*, **48**(4), pp. 1413–1422, 2013. DOI: 10.1007/s10853-012-6892-2; *J. Mater. Sci.*, **55**(21), p. 9144, 2020.

[15] Mahmoudi, M., Elwany, A., Yadollahi, A., Thompson, S.M., Bian, L. & Shamsaei, N., Mechanical properties and microstructural characterization of selective laser melted 17-4 PH stainless steel. *Rapid Prototyp. J.*, 2017.

[16] Maconachie, T. et al., SLM lattice structures: Properties, performance, applications and challenges. *Mater. Des.*, **183**, 2019.

[17] Köhnen, P., Haase, C., Bültmann, J., Ziegler, S., Schleifenbaum, J.H. & Bleck, W., Mechanical properties and deformation behavior of additively manufactured lattice structures of stainless steel. *Mater. Des.*, **145**, pp. 205–217, 2018.

[18] Concli, F., Gilioli, A. & Nalli, F., Experimental – numerical assessment of ductile failure of additive manufacturing selective laser melting reticular structures made of Al A357. *Proc. Inst. Mech. Eng. Part C: J. Mech. Eng. Sci.*, **32**(7), pp. 3047–3056, Mar. 2019.

[19] Nalli, F., Cortese, L. & Concli, F., Ductile damage assessment of Ti6Al4V, 17-4PH and AlSi10Mg for additive manufacturing. *Eng. Fract. Mech.*, **241**, 2021.

[20] Bonaiti, L., Concli, F., Gorla, C. & Rosa, F., Bending fatigue behaviour of 17-4 PH gears produced via selective laser melting. *Procedia Struct. Integr.*, **24**, pp. 764–774, 2019.

[21] Concli, F. & Gilioli, A., Numerical and experimental assessment of the static behavior of 3D printed reticular Al structures produced by selective laser melting: Progressive damage and failure. *Procedia Struct. Integr.*, **12**, pp. 204–212, 2018.

[22] Ramberg, W. & Osgood, W.R., *Description of Stress–Strain Curves by Three Parameters*, Washington DC, 1943.

[23] Hollomon, J.R., Tensile deformation. *Trans. AIME*, **162**, pp. 268–277, 1945.

[24] Maccioni, L., Rampazzo, E., Nalli, F., Borgianni, Y. & Concli, F., Low-cycle-fatigue properties of a 17-4 PH stainless steel manufactured via selective laser melting. *Material and Manufacturing Technology XI*, vol. 877, pp. 55–60, 2021.

[25] Maccioni, L., Fraccaroli, L. & Concli, F., High-cycle-fatigue characterization of an additive manufacturing 17-4 PH stainless steel. *Key Engineering Materials*, 2020.

[26] ASTM, Standard practice for statistical analysis of linear or linearized stress-life (s–n) and strain-life (e–n) fatigue data. *E ASTM 739-91*, 2006.

[27] Maccioni, L., Fraccaroli, L., Borgianni, Y. & Concli, F., High-cycle-fatigue characterization of an additive manufacturing 17-4 PH stainless steel. *11th International Conference on Materials and Manufacturing Technologies*, p. MT013, 2020.

[28] Alomar, Z. & Concli, F., A review of the selective laser melting lattice structures and their numerical models. *Adv. Eng. Mater.*, **22**(12), 2020.

[29] www.code-aster.org.

[30] Concli, F. & Gilioli, A., Numerical and experimental assessment of the mechanical properties of 3D printed 18-Ni300 steel trabecular structures produced by selective laser melting – A lean design approach. *Virtual Phys. Prototyp.*, 2019.

[31] https://dakota.sandia.gov/.

DESIGN FOR ADDITIVE MANUFACTURING: IS IT AN EFFECTIVE ALTERNATIVE? PART 2 – COST EVALUATION

FRANCO CONCLI[1], MARGHERITA MOLINARO[2] & ELEONORA RAMPAZZO[1]
[1]Free University of Bolzano/Bozen, Faculty of Science and Technology, Italy
[2]University of Udine, Polytechnic Department of Engineering and Architecture, Italy

ABSTRACT

Additive Manufacturing (AM) is becoming a more and more widespread (and trendy) approach. Its flexibility and capability to manufacture any topology has opened new possibilities: AM could lead to significant performance improvements thanks to the exploitation of lattice or reticular structures as partial replacement of the traditional solid design. The potential of this technology knows no bounds. However, in the real world, the lower performances of the materials and the high manufacturing costs significantly restrict the fields of application for which the adoption of AM results effective. In this context, the mechanical static and fatigue properties of a 17-4 PH Stainless Steel produced via AM were experimentally measured and compared with those of the wrought material to quantify the performance reduction. Based on these data, three components, namely a hip prosthesis, a blow plastic bottle die, and an automotive gear were selected as representative examples to show the pros and contra of AM. The three components were chosen because they belong to three quite dissimilar fields and are produced in different batch sizes. The three original designs were specifically optimized for AM by means of Finite Element (FE) Simulations. The new solutions fulfil the strength requirements of the original parts showing at the same time reduced weights and inertias. The traditional and new designs were compared in terms of production times and costs to quantify the real benefits of AM for different applications.

Keywords: additive manufacturing, FEM, optimization.

1 INTRODUCTION

Based on the optimization of the design carried out in the first part of this work, in this second paper an economical/feasibility analysis of the optimized solutions for each of the analysed case studies ("Hip Prosthesis", "Bottle Blowing Mold" and "Automotive Gear") is made. To better highlight drawbacks and benefits of the Additive Manufacturing (AM), three solutions for each case study were analysed. Specifically, the original design was virtually manufactured with traditional operations, relying on the Computer Numerical Control (CNC) machine, and with Selective Laser Melting (SLM) machine. These two solutions were compared with the optimized counterparts relying on a lattice internal structure, manufactured with the SLM machine.

The outcomes of the analysis are aimed at showing the effectiveness of the new technology, both in terms of production costs and times.

2 COST CALCULATION PROCEDURE

In the literature, different costing techniques, as well as different cost drivers and areas of applications, have been used to estimate unit costs [1], [2]. Material costs, labor costs and overhead costs, which include all cost elements other than the previous two, are typically considered in the cost calculation techniques [3], [4]. What differs among the various models proposed in the literature is not only the approach adopted for the calculation (i.e., task-based vs. level-based), but also the number and type of cost drivers, beyond labor and material costs, considered in the analysis (e.g., machine, electricity, set-up, tooling, inventory, logistics costs etc.) [1].

In line with other studies comparing total costs in traditional and additive manufacturing environments [5]–[7], in this study a task-based approach using a process-oriented cost model was adopted. Thus, only the production process was included in the calculations. Moreover, pre-processing and part manufacturing were considered separately. These phases were selected to properly represent different cost centers, thus facilitating the calculation and making it easier to apply the model in other contexts [8].

Each phase was associated to a certain number of cost drivers. The pre-processing phase, which deals with all the activities that precede the effective production, included all the costs for the preparation of CAM software. The part manufacturing phase, which represents the effective production, included instead all costs related to material, labor, machines, tools, and electricity. The calculation of these cost drivers varied according to the manufacturing type.

As the above overview shows, the model did not aim to calculate the effective total cost of the two solutions, but only the sum of cost drivers that differ between them. Accordingly, only the factors directly affecting the part cost were considered, in line with [5] and [6]. All the other costs, such as administrative overhead, logistics, rental costs, etc. [9], [10], can be considered invariant among traditional and additive manufacturing environments and, therefore, they were not taken into account in the model. Furthermore, as highlighted in previous studies [10], [11], the abovementioned terms have an effect on the total cost limited to 10%. Therefore, their inclusion in the calculations would not significantly affect the results.

Before providing a detailed description of the cost drivers, it is worth clarifying the assumptions at the basis of our calculation model. First of all, it was assumed that the manufacturing plant works 16 hours per day, 5 days per week and 48 weeks per year, with a consequent total productive number of hours per year equal to 3,840 h/year. Second, it was assumed that the manufacturing plant is located in Italy and thus this country was used as a reference for all the estimations. In particular, we selected a medium-sized Italian company operating in the engineering sector to collect reliable information on cost drivers (e.g., hourly costs of labor and electricity, material costs, tools costs, etc.), as well as to estimate the time needed for the various activities (e.g., set-up time). Finally, the straight-line depreciation technique was adopted to calculate the machine hourly cost, in line with [5], [6], [9]. This required the estimation of the total cost and the economic life of the production machines. Dividing the total cost by the number of useful life hours, the machine cost per hour, which was taken into consideration for both set-up and production activities, could be derived.

The following paragraphs explain how the costs were estimated in the two manufacturing environments.

2.1 Cost calculation for traditional manufacturing

In traditional manufacturing, the production consists of two main phases processed on the same CNC machine. The first phase (i.e., roughing) subtracts the waste material from an initial rough block volume of steel to obtain the rough shape of the product. Starting from this, the second phase (i.e., surface finishing) finishes the piece by removing a further layer of material to obtain the final required quality of the product.

The six cost drivers used for the calculation of the total cost in this manufacturing environment are shown and explained in Table 1.

The preparation of CAM software represents the first cost item, calculated multiplying the hourly cost of the programming by the time needed for such activity. The hourly cost of the programming was obtained considering both software cost and labor cost of programming, while the working time was estimated taking into consideration that both roughing and surface finishing phases require the development of a customized program.

Table 1: Cost calculation procedure for traditional manufacturing.

CAM programming cost	Hourly cost for CAM programming	€/h	PC
	Programming time	h	PT
	Total CAM programming cost	€	CAM = PC·PT
Material cost	Rough block volume	mm^3	V
	Rough block weight	kg	W = 0.008/1,000·V
	Unit material cost	€/kg	UC
	Total material cost	€/pc	MAT = W*UC
Labor cost	CNC manufacturing time	min/pc	MT
	Worker time for manufacturing	min/pc	WT = 0.1·MT
	Worker time for machine set-up	min/pc	ST
	Hourly labor cost	€/h	LC
	Total labor cost	€/pc	LAB = (WT + ST)·LC/60
Tool cost	Tool useful life	min	TL
	Unit tool cost	€/tool	TC
	Total tool cost	€/pc	TOO = (MT/L)·TC
CNC machine cost	Working hours per year	h/year	H
	CNC machine useful life	year	ML
	Total CNC machine cost	€	MC
	Total CNC machine cost	€/pc	MCC = (ST + MT) · (MC/H/ML/60)
Energy cost	CNC consumed electrical power	kW	P1
	Hourly energy cost	€/kWh	EC
	Total energy cost	€/pc	ENE = (MT/60)·P1·EC
Total cost	Production volume	pc	N
	Total unit cost	€/pc	**C = CAM + N· (MAT + LAB + TOO + MCC + ENE)**

The material cost was instead computed considering the initial rough block volume of steel, from which to subtract the waste material. The dimensions of the initial block were chosen according to the object to be manufactured (i.e., hip prosthesis, bottle blowing mold or automotive gear) and its weight was estimated by supposing a density of 8 g/cm^3. Multiplying the weight of the block by the cost of a unit weight of material, provided by a supplier of the reference company, the total material cost was obtained.

For the labor cost, the time needed for the operator to monitor the machine (i.e., manufacturing activity) and that needed to load and unload the pieces from the machine (i.e., set-up activity) was taken into account in the calculations. The former was defined as the 10% of the total manufacturing time and the latter was estimated with the support of the reference company, which also provided the data of the hourly labor cost. The total labor cost was simply given by the product between the total working time of the operator and the hourly labor cost.

The traditional CNC machine requires a tool for the manufacturing. After the estimation of unit tool cost and tool useful life (using the Taylor's formula), the total tool cost was calculated multiplying the number of tools needed for the manufacturing of one single piece by the unit tool cost.

The machine cost was calculated considering the money spent to buy the machine, its useful life and the expected working hours per year, whose product provided the hourly machine cost. This latter cost, multiplied by the total time during which the machine is expected to be employed, provided an estimation of total machine cost per piece.

Finally, to estimate the energy cost, electrical power consumption was multiplied by the hourly energy cost, provided again by the reference company.

Obviously, to calculate the total cost of a single piece, variable and fixed costs were distinguished. The only fixed cost item in our analysis was represented by the programming cost, whose value is independent from the number of manufactured units. Thus, the total cost was calculated by multiplying all variable costs by the hypothesized volume and summing the result to the CAM programming cost.

For a more complete understanding of the calculation process, an overview of how the total manufacturing time was calculated is shown in Table 2. In particular, the table distinguishes between roughing and surface finishing phases. Each of them has different waste volumes (i.e., volumes to be subtracted from the block) and different working speeds. By summing the time needed to carry out each phase, the total manufacturing time, namely the time during which a single piece is processed on the CNC machine, was obtained.

Table 2: Calculation procedure for manufacturing time.

Roughing data	Waste volume roughing	mm³	S
	MRR roughing	mm³/min	MRRS
Surface finishing data	Waste volume surface finishing	mm³	F
	MRR surface finishing	mm³/min	MRRF
	CNC manufacturing time	min/pc	MT = S/MRRS + F/MRRF

2.2 Cost calculation for AM

In AM, the production consists of two main phases, processed on different machines. In the first phase, the SLM machine is used, and the rough shape of the product is obtained. As in traditional manufacturing, this phase is followed by a surface finishing, carried out with a CNC machine, which finishes the piece by filing a further layer of material to obtain the final configuration of the product.

The cost items used for the calculation are reported in Table 3. The procedure was similar to the one described for traditional manufacturing, but with some differences.

First of all, the machine cost was calculated for both CNC and SLM machines. For what concerns this latter, the speed time was estimated by checking several values provided by some SLM equipment suppliers regarding machines with specifications similar to the one used for the 17-4 PH SS samples. Starting from this value and considering the volume to be produced, the total manufacturing time on SLM machine could be calculated. Finally, the overall machine cost was estimated considering, as for traditional manufacturing, the money spent to buy the machine, its useful life, and the expected working hours per year.

Table 3: Cost calculation procedure for AM.

CAM programming cost	Hourly cost for CAM programming	€/h	PC
	Programming time	h	PT
	Total CAM programming cost	€	CAM = PC*PT
Material cost	Rough piece volume	mm³	V
	Rough piece weight	kg	W = 0.008/1,000·V
	Unit material cost	€/kg	UC
	Total material cost	€/pc	MAT = W*UC
Labor cost	CNC manufacturing time	min/pc	MT
	Worker time for manufacturing	min/pc	WT = 0.1*MT
	Worker time for machine set-up	min/pc	ST
	Hourly labor cost	€/h	LC
	Total labor cost	€/pc	LAB = (WT + ST)·LC/60
Tool cost	Tool useful life	Min	TL
	Unit tool cost	€/tool	TC
	Total tool cost	€/pc	TOO = (MT/L)·TC
CNC machine cost	Working hours per year	h/year	H
	CNC machine useful life	Year	ML
	Total CNC machine cost	€	MC
	Total CNC machine cost	€/pc	MCC = (ST + MT)·(MC/H/ML/60)
SLM machine cost	Speed SLM production	cm³/h	SP
	SLM manufacturing time	min/pc	M = V/1,000/SP·60
	SLM machine useful life	Year	SL
	Total SLM machine cost	€	SC
	Total SLM machine cost	€/pc	MSC = M·(SC/H/SL/60)
Energy cost	CNC consumed electrical power	kW	P1
	SLM consumed electrical power	kW	P2
	Hourly energy cost	€/kWh	EC
	Total energy cost	€/pc	ENE = [(MT/60)·P1 + (M/60)·P2]·EC
Total cost	Production volume	pc	N
	Total unit cost	€/pc	**C = CAM + N*(MAT + LAB + TOO + MCC + MSC + ENE)**

For what concerns instead all the other cost items, the differences compared to traditional manufacturing included:

- the programming time, which was defined taking into consideration that only the surface finishing phase requires a customized program, while no programming is needed for SLM machine;
- the rough piece volume, which was equal to the final piece volume plus the small waste generated in the surface finishing phase;

- the CNC manufacturing time, which included only the time needed for the surface finishing phase;
- the worker time for machine set-up, which included the time needed to move the pieces from the SLM to the CNC machine and to unload the final piece from this latter at the end of the production; no labor was considered for the SLM production; and
- the total energy cost, which was calculated considering both SLM and CNC energy consumptions, as shown in Table 3.

3 COST ANALYSIS AND RESULTS

Using the cost models described in the previous section, we estimated the unit cost of the three products (i.e., hip prosthesis, bottle blowing mold or automotive gear) in three potential situations: use of traditional manufacturing, use of AM and use of AM with geometry optimization through lattice internal structure. The main cost advantages of this latter case are the need of a lower volume of material, which results into a lower material cost, and the consequent reduction of SLM manufacturing time, which in turn decreases the machine cost for the product.

The total unit cost of bottle blowing mold and automotive gear was also distinguished for different batch volumes. Indeed, while the hip prosthesis can be considered a highly customized product manufactured in single batches, the other two components are typically produced with more numerous volumes. In particular, a batch of 10 pieces was considered for the bottle blowing mold and one of 1,000 pieces was hypothesized for the automotive gear. As we previously highlighted, a more numerous batch allows to reduce the unit cost of programming.

The manufacturing time for the CNC machine required the estimation of the Material Removal Rate (MRR) and waste volumes. Relying on the reference company's data, the MRR roughing was set at 72,000 mm³/min for hip prosthesis and bottle blowing mold and at 18,000 mm³/min for automotive gear. The MRR surface finishing was instead estimated as 1,800 mm³/min for hip prosthesis and bottle blowing mold and as 450 mm³/min for automotive gear. The waste volumes were calculated considering the difference between rough block/piece volumes and the final expected volume of each product. They are shown in Table 4, together with the resulting CNC manufacturing times. These latter were calculated with the formula reported in Table 2 and considering that, while in traditional manufacturing the CNC machine is used for both roughing and surface finishing phases, with significant waste volumes, in AM the CNC machine is employed only for surface finishing.

Table 4: Waste volumes and CNC manufacturing time.

		Hip prosthesis	Bottle blowing mold	Automotive gear
Waste volume roughing	mm³	381,051.64	1,635,734.08	22,835.51
Waste volume surface finishing	mm³	778.97	81,796.37	1,082.67
CNC manufacturing time for traditional manufacturing	min/pc	5.73	68.16	3.67
CNC manufacturing time for additive manufacturing	min/pc	0.43	45.44	2.41

The detailed results of the cost calculations for traditional manufacturing and AM are shown in Tables 5 and 6 respectively. A summary of the results for the three components is further depicted in Figs 1–3. For what concerns AM with geometry optimization, the calculation of the rough piece volume was carried out by considering the results obtained in the optimization activity in terms of average cell volumes and cell box volumes. These values were equal to 28 mm^3 and 35.9 mm^3 respectively.

Starting from Tables 5 and 6 data, we also carried out some sensitivity analysis. In particular, we recalculated the total unit cost of the three products first by modifying machines purchase costs and then unit material cost. The results were similar to those shown in the tables, giving support to their reliability.

For a more complete analyses of the various manufacturing solutions, we also compared the time needed to produce a single hip prosthesis, a bottle blowing mold and an automotive gear, distinguishing between different batch volumes (V). The results are shown in Table 7.

Table 5: Total unit cost for traditional manufacturing.

		Hip prosthesis	Bottle blowing mold		Automotive gear	
Hourly cost for CAM programming	€/h	40	40		40	
Programming time	h	4	8		1	
Total CAM programming cost	€	160	320		40	
Rough block volume	mm^3	420,000	5,725,552.61		76,969.02	
Rough block weight	kg	3.36	45.80		0.62	
Unit material cost	€/kg	3	3		3	
Total material cost	€/pc	10.08	137.41		1.85	
CNC manufacturing time	min/pc	5.73	68.16		3.67	
Worker time for manufacturing	min/pc	0.57	6.82		0.37	
Worker time for machine set-up	min/pc	5	5		3	
Hourly labor cost	€/h	20	20		20	
Total labor cost	€/pc	1.86	3.94		1.22	
Tool useful life	min	35	35		90	
Unit tool cost	€/tool	100	100		200	
Total tool cost	€/pc	16.36	194.75		8.17	
Working hours per year	h/year	3,840	3,840		3,840	
CNC machine useful life	year	15	15		15	
Total CNC machine cost	€	400,000	400,000		400,000	
Total CNC machine cost	€/pc	1.24	8.47		0.77	
CNC consumed electrical power	kW	6.62	6.62		1.65	
Hourly energy cost	€/kWh	0.27	0.27		0.27	
Total energy cost	€/pc	0.17	2.03		0.03	
Production volume	pc	1	1	10	1	100
Total unit cost	€/pc	**189.71**	**666.59**	**378.59**	**51.94**	**11.98**

Table 6: Total unit cost for AM with and without geometry optimization.

		Hip prosthesis	Bottle blowing mold	Automotive gear
Hourly cost for CAM programming	€/h	40	40	40
Programming time	h	2	4	0.5
Total CAM programming cost	€	80	160	20
Rough piece volume	mm³	38,948.36 (30,548.97)	4,089,818.53 (3,207,830.37)	54,133.51 (42,459.37)
Rough piece weight	kg	0.31 (0.24)	32.72 (25.66)	0.43 (0.34)
Unit material cost	€/kg	3	3	3
Total material cost	€/pc	0.93 (0.73)	98.16 (76.99)	1.30 (1.02)
CNC manufacturing time	min/pc	0.43	45.44	2.41
Worker time for manufacturing	min/pc	0.04	4.54	0.24
Worker time for machine set-up	min/pc	1	1	0.5
Hourly labor cost	€/h	20	20	20
Total labor cost	€/pc	0.35	1.85	0.25
Tool useful life	min	35	35	35
Unit tool cost	€/tool	100	100	100
Total tool cost	€/pc	1.24	129.84	6.87
Working hours per year	h/year	3,840	3,840	3,840
CNC machine useful life	year	15	15	15
Total CNC machine cost	€	400,000	400,000	400,000
Total CNC machine cost	€/pc	0.17	5.38	0.34
Speed SLM production	cm³/h	25	25	25
SLM manufacturing time	min/pc	93.48 (73.32)	9,815.56 (7,698.79)	129.92 (101.90)
SLM machine useful life	year	8	8	8
Total SLM machine cost	€	500,000	500,000	500,000
Total SLM machine cost	€/pc	25.36 (19.89)	2,662.64 (2,088.43)	35.24 (27.64)
CNC consumed electrical power	kW	0.16	0.16	0.04
SLM consumed electrical power	kW	0.2	0.2	0.2
Hourly energy cost	€/kWh	0.27	0.27	0.27
Total energy cost	€/pc	0.08 (0.07)	8.87 (6.96)	0.12 (0.09)
Production volume	pc	1	1	10
Total unit cost	€/pc	**108.13 (102.44)**	**3,066.72 (2,469.44)**	**2,922.72 (2,325.44)**

Note: Numbers in parentheses represent the values associated to the optimized geometry; they are specified only when they differ from those obtained for the manufacturing without geometry optimization.

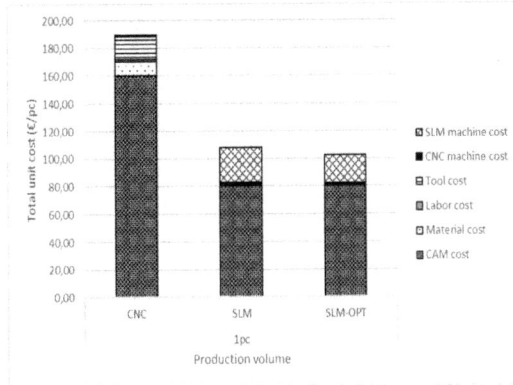

Figure 1: Total unit cost for hip prosthesis.

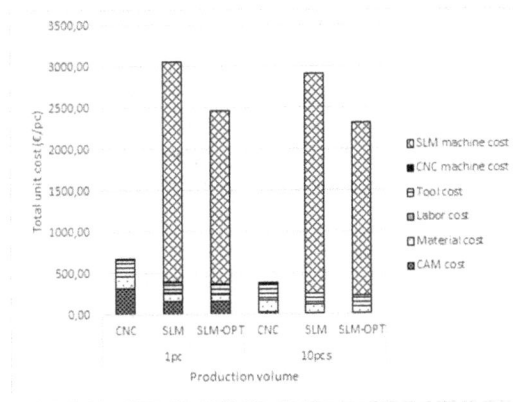

Figure 2: Total unit cost for bottle blowing mold at different batch sizes.

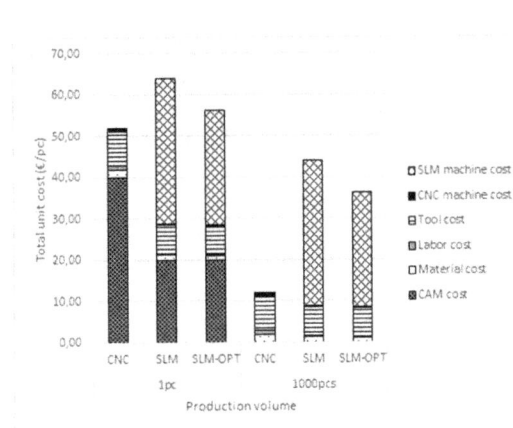

Figure 3: Total unit cost for automotive gear at different batch sizes.

Table 7: Total manufacturing time (min) for a single piece production.

| | Hip prosthesis | Bottle blowing mold | | Automotive gear | |
	V = 1	V = 1	V = 10	V = 1	V = 1,000
CNC	250.73	553.16	121.16	66.67	6.73
SLM	214.91	10,102.01	9,886	162.83	132.86
SLM-OPT	194.75	7,985.24	7,769.24	134.81	104.84

Note: SLM-OPT refers to the use of AM with optimized geometry.

4 DISCUSSION

An analysis of the results shown in Tables 5 and 6 allows to make several observations. First of all, the CAM programming cost is significantly higher in traditional manufacturing. Therefore, in this context, the cost of a single piece is significantly influenced by the cost of programming, whose value reaches 84%, 48% and 77% respectively for hip prosthesis, bottle blowing mold and automotive gear. Obviously, when higher production volumes are hypothesized, this cost is spread among numerous pieces, making its effect on the total unit cost almost irrelevant. The result is that an increase of the production volume reduces the unit product cost in traditional manufacturing more than what happens in AM. Not by change, the traditional manufacturing becomes more and more convenient when numerous batches are considered. However, in general, we cannot state that the convenience of AM or not to AM strictly depends on the number of pieces to be produced, since three different situations emerge from the three product cases.

As for the prosthesis, AM appears the most convenient manufacturing solution because the programming cost of the CNC machine is very high, if compared to the other cost items.

For what concerns the bottle blowing mold, the amount of material needed for production and the resulting long SLM manufacturing times make this solution much more expensive than traditional manufacturing. The cost of the SLM machine is even 86% of the total unit cost of the product. This cost difference between the two solutions becomes even more significant as the production volume increases.

Finally, as regards the automotive gear, an intermediate situation emerges. The SLM machine cost is very high, but so does the CNC programming cost. Therefore, even if the latter is more convenient, the cost difference is rather limited, if compared to that of bottle blowing mold. However, by bringing the production volume to 1,000 pieces, the unit SLM machine cost remains the same, while that of CNC programming is significantly reduced, reaching an incidence lower than 1%. Traditional manufacturing becomes therefore extremely convenient in this context.

Overall, making a purely economic evaluation, the analysis seems to suggest that the convenience of AM or not AM depends not only on the number of pieces, but also on the shape and size of the pieces to be produced. The hip prosthesis has a very small product volume: consequently, SLM machine cost and material purchase cost are not so high. The CNC programming cost is instead quite considerable, making the traditional manufacturing solution less convenient than the AM one. The bottle blowing mold has instead a very large product volume, which significantly increases the 3D printing times and, consequently, also the SLM machine cost. This latter, in particular, exceeds the programming costs of traditional manufacturing, making AM less convenient.

Obviously, this economic evaluation should be accompanied an analysis of the total time needed for manufacturing (see Table 7). As it could be expected, the total production times of bottle blowing molds and automotive gears are much higher in AM, especially when the volumes increase. Surprisingly, the production time of the hip prosthesis is instead shorter in the AM case. The reason for this result is similar to that proposed for the cost evaluation. The hip prosthesis requires indeed a significant CNC machine programming time, which accounts for a relevant part in the total manufacturing time.

5 CONCLUSIONS

This paper investigated the effectiveness of AM as alternative to traditional manufacturing by comparing mechanical properties as well as production costs and times of three components produced in the two manufacturing environments. The original designs of the three components were also optimized for AM by means of Finite Elements Simulations.

From the analyses, it emerged that the mechanical performances of additive-produced materials are comparable to those produced with traditional manufacturing. This element is therefore not particularly discriminating in the choice of one or the other technology. However, it is also true that AM has the advantage of being extremely versatile, allowing the creation of structures not producible otherwise. This opens the way for extreme optimization, such as that proposed in the first part of the paper.

Some differences exist instead in terms of production times and costs. In general, traditional techniques have higher fixed costs and shorter production times; they seem therefore more suitable for large batches. However, the analysis of components with very different structures suggested also that there is no general rule for the choice. For instance, contrary to all the expectations, the use of AM resulted to be less convenient for a bottle blowing mold, which has an average number of elements, than for automotive gear, whose elements are more numerous. The explanation lies in the fact that the mold has a simple geometry and a very high volume, which makes AM extremely slow and expensive.

Overall, it is possible to conclude that, to choose between the two alternative technologies, it is always necessary to analyze the specific characteristics of the item to be produced, in terms of shape, volume and structure. However, it is also worth highlighting that AM allows to create unique designs ensuring, for example, significant weight reductions or better weight distributions. Moreover, the lattice structures could be exploited to modify the heat transfer capability or to shift the eigen frequencies of the systems and, consequently, to improve the NVH (Noise, Vibration and Harshness) behavior. In this sense, if the design is optimized for AM, this new technology could really make the difference.

REFERENCES

[1] Kadir, A.Z.A., Yusof, Y. & Wahab, M.S., Additive manufacturing cost estimation models: A classification review. *The International Journal of Advanced Manufacturing Technology*, **107**(9–10), pp. 4033–4053, 2020.

[2] Costabile, G., Fera, M., Fruggiero, F., Lambiase, A. & Pham, D., Cost models of additive manufacturing: A literature review. *International Journal of Industrial Engineering Computations*, **8**(2), pp. 263–282, 2016.

[3] H'mida, F., Martin, P. & Vernadat, F., Cost estimation in mechanical production: The cost entity approach applied to integrated product engineering. *International Journal of Production Economics*, **103**(1), pp. 17–35, 2006.

[4] Mandolini, M., Campi, F., Favi, C., Germani, M. & Raffaeli, R., A framework for analytical cost estimation of mechanical components based on manufacturing knowledge representation. *The International Journal of Advanced Manufacturing Technology*, **107**(3–4), pp. 1131–1151, 2020.

[5]   Atzeni, E., Iuliano, L., Minetola, P. & Salmi, A., Redesign and cost estimation of rapid manufactured plastic parts. *Rapid Prototyping Journal*, **16**(5), pp. 308–317, 2010.

[6]   Atzeni, E. & Salmi, A., Economics of additive manufacturing for end-usable metal parts. *The International Journal of Advanced Manufacturing Technology*, **62**(9–12), pp. 1147–1155, 2012.

[7]   Hällgren, S., Pejryd, L. & Ekengren, J., Additive manufacturing and high speed machining cost comparison of short lead time manufacturing methods. *Procedia CIRP*, **50**, pp. 384–389, 2016.

[8]   Lindemann, C., Jahnke, U., Moi, M., & Koch, R., Analyzing product lifecycle costs for a better understanding of cost drivers in additive manufacturing. in *23rd Annual International Solid Freeform Fabrication Symposium – An Additive Manufacturing Conference, SFF 2012*, pp. 177–188, 2012.

[9]   Özbayrak, M., Akgün, M., & Türker, A.K., Activity-based cost estimation in a push/pull advanced manufacturing system. *International Journal of Production Economics*, **87**(1), pp. 49–65, 2004.

[10]   Ruffo, M., Tuck, C., & Hague, R., Cost estimation for rapid manufacturing – Laser sintering production for low to medium volumes. *Proceedings of the Institution of Mechanical Engineers, Part B: Journal of Engineering Manufacture*, **220**(9), pp. 1417–1427, 2006.

[11]   Ruffo, M., Tuck, C., & Hague, R., Make or buy analysis for rapid manufacturing. *Rapid Prototyping Journal*, **13**(1), pp. 23–29, 2007.

RELIABLE GEAR DESIGN: TRANSLATION OF THE RESULTS OF SINGLE TOOTH BENDING FATIGUE TESTS THROUGH THE COMBINATION OF NUMERICAL SIMULATIONS AND FATIGUE CRITERIA

FRANCO CONCLI[1], LORENZO MACCIONI[1] & LUCA BONAITI[2]
[1]Faculty of Science and Technology, Free University of Bolzano, Italy
[2]Department of Mechanical Engineering, Politecnico di Milano, Italy

ABSTRACT

Establishing the actual gear root bending strength is a fundamental aspect in gear design. With this respect, gears materials can be characterized through two types of tests, i.e. on Running Gears (RG) or Single Tooth Bending Fatigue (STBF). The former is able to reproduce the loading conditions of the actual gears and, therefore, leads to the most accurate results. The latter excels in terms of efficiency and simplicity of the experimental campaign but as a drawback, tends usually to overestimate the material strength due to the different stress state histories it induces on the tooth root. Therefore, a common practice is to carry out STBF tests and apply a correction coefficient (f_{korr}) for exploiting the results in the design of actual gears. In the present paper, an approach to estimate f_{korr} centered on the combination of numerical simulations and multi-axial fatigue criteria based on the critical plane capable of taking into account non-proportional loading conditions has been proposed. In particular, the same gear geometry has been simulated through Finite Element (FE) models in two conditions, i.e. STBF and RG. The outcomes of the simulations, in terms of stress histories in the tooth root region, have been analyzed with five different fatigue criteria, i.e. Findley, Matake, McDiarmid, Papadopoulos, and Susmel et al. f_{korr} has been calculated as the ratio between the maximum damage parameter observed in the STBF and RG conditions according to the different fatigue criteria. Results show that f_{korr}, calculated for three different materials (i.e. 18NiCrMo5, 42CrMoS4, 31CrMo12), differs up to 22% between the RG and the STBF conditions (depending on the criterion considered). Therefore, future studies should aim to understand which fatigue criterion is the most appropriate for this type of analysis.
Keywords: material characterization, STBF, FEM, gears, multiaxial fatigue, critical plane.

1 INTRODUCTION

Through the meshing of teeth with a conjugate profile, gears transfer mechanical power, in terms of torque and rotational speed, between two non-coaxial rotating shafts [1]. Nevertheless, the repeated sliding/rolling contact between the tooth flanks could lead to different fatigue failure modes [2]. For instance, high contact pressure can lead to damage in the meshing area such as wear, scuffing, pitting, and micro-pitting [1]. However, the most dangerous failure mode is due to the tooth root bending fatigue [3], [4]. Indeed, the transmitted forces during the meshing induce varying stresses on the tooth root fillet region that, in turn, potentially lead to cracks nucleation and propagation to the tooth root breakage [5].

In gear design, the prevention of the failure due to the tooth root bending fatigue is a primary objective [6] and it is supported by standards, e.g. (ISO 6336-3 [7] and ANSI/AGMA, 2001 [8]). To determine the load carrying capacity of a gear, standards suggest verifying (through specific calculation method) that the maximum stress σ_F at the tooth root due to pure bending does not exceeds the permissible bending stress σ_{FP}. According to the Method B of ISO 6336-3 [7], σ_{FP} is proportional to material strength σ_{Flim} that, in turn, is usually determined though experimental campaigns.

WIT Transactions on Engineering Sciences, Vol 130, © 2021 WIT Press
www.witpress.com, ISSN 1743-3533 (on-line)
doi:10.2495/CMEM210101

To characterize σ_{Flim} two types of tests can be carried out, i.e. tests on Running Gears (RG) [6], [9] and tests on Single Tooth Bending Fatigue (STBF) [10]–[13].

In RG tests, test rigs have to be able to provide a given torque with proper lubrication to engaging RG [8], [14]. The specimen consists of a gear made of the material (and the treatments/finishes) to be tested. The specimen engages with one or more gears and, when a tooth breaks, the number of times the tooth has meshed is noted, and the entire specimen is replaced. Therefore, RG tests are able to reproduce the exact stress state of the actual gears and allow obtaining reliable value of σ_{Flim} [15]. However, as a drawback, the experimental campaign results particularly long and expensive.

As for RG tests, also in STBF tests the specimen consists of a gear made of the material to be tested. In these configuration, pulsating forces are applied to two teeth of the same specimen through two anvils. Exploiting the Wildhaber property, these forces are tangent to the base circumference and normal to the tested teeth flanks. STBF tests can be performed on universal testing machine (which does not require lubrication of the sample) and multiple tests on a single specimen can be carried out since the forces are applied to two teeth per test. As a drawback, experimental evidences have shown that the results of STBF tests tend to overestimate the value of σ_{Flim} [16], [17]. This is due to the different stress histories that RG and STBF induces on the tooth root witch were neglected by the standards [18].

Indeed, the stress histories at the tooth root differ for the following reasons. First, in STBF tests, for keeping the specimen in the right place during the test, a compressive load should be always present; the typical ratio between the minimum and maximum force applied to the teeth is R = 0.1, e.g. the STBF tests conducted in [14], [19]–[23]. Naturally, this differs from RG where R = 0 and, therefore, it modifies the mean stress [9], [11], [24], [25]. Second, in STBF tests, the forces are applied with a fixed direction and position and vary in a sinusoidal way with a constant amplitude while, in RG tests, both the force magnitude and direction are variable. In addition, the force direction in STBF tests can be different from the one in the Outer Point of Single pair tooth Contact (OPSC) of RG and, therefore, a different share between pure bending and pure compressive stresses could be present. Moreover, in RG tests, the variable number of mating teeth pairs leads to an uneven force sharing [26]. Consequently, the stress time history at the tooth root is not sinusoidal as in the STBF tests [9], [11].

To compensate these effects, a correction coefficient (f_{korr}) can be exploited. f_{korr} is representative of the ratio between the σ_{Flim} obtained via STBF and the σ_{Flim} obtained via RG (when the two tests were set to produce the same σ_F according to the standard (ISO 6336-3)). This coefficient has been estimated experimentally by Rettig [16] and Stahl [17]. They proposed to exploit a constant value of $f_{korr} = 0.9$. However, tests were conducted for a limited combination of materials and geometries and, therefore, new techniques to estimate f_{korr} for each specific combination of material and geometry have been developed. An advanced method is presented in [26], where the scholars have combined the Crossland fatigue criterion [27] with a numerical simulation of RG and the experimental results of STBF tests. In this case, the corrective factors have resulted equal to 0.82 and 0.84 for the two materials tested. In the present paper, an innovative approach to calculate f_{korr} through the combination of numerical simulations and fatigue criteria capable to take into account the different stress histories emerging in RG and STBF tests has been proposed. In particular, the main fatigue criteria based on the critical plane have been considered in this work. This approach has been implemented on a gear geometry previously tested by the authors' research group [28].

2 BACKGROUND

In the present section, an overview of the main fatigue criteria based on critical plane (i.e. Findley [29], Matake [30], McDiarmid [31], Papadopoulos [32], and Susmel et al. [33] exploited in this work is presented.

Having the stress histories (in term of stress tensors) referred to a specific point (eqn (1)) it is possible to evaluate the maximum octahedral stress $\sigma_{h,max}$ (in the time window T) according to eqn (2). Where $\boldsymbol{\sigma_O}$ is a vector containing the principal stresses that, for the same time instant t, satisfies eqn (3), the relation, where $\overline{\overline{I}}$ is the identity matrix.

In addition, it is possible to calculate the stress vector $\boldsymbol{P_n}$ acting on a plane defined by a normal vector $\boldsymbol{n}(\phi_n, \theta_n)$ through the relation showed in eqn (4). The modulus and the direction of $\boldsymbol{P_n}$ vary in time (Fig. 1(a)). In addition, $\boldsymbol{P_n}$ can be decomposed into a normal component $\boldsymbol{\sigma_n}$ (eqn (5)), having time-varying modulus and fixed direction, and a tangential component $\boldsymbol{\tau_n}$, having time-varying modulus and direction that, in turn, can be decomposed in its component aligned with the $\boldsymbol{u}$ and $\boldsymbol{v}$ directions (eqn (6)) (Fig. 1(b)). Where $\boldsymbol{n}, \boldsymbol{u}, \boldsymbol{v}$ are defined in eqn (7).

$$\overline{\overline{\sigma}}(t) = \begin{bmatrix} \sigma_{xx}(t) & \tau_{xy}(t) & \tau_{xz}(t) \\ \tau_{yx}(t) & \sigma_{yy}(t) & \tau_{yz}(t) \\ \tau_{zx}(t) & \tau_{zy}(t) & \sigma_{zz}(t) \end{bmatrix}, \tag{1}$$

$$\sigma_{h,max} = \max_T \left\{ \frac{1}{3} \Sigma_{i=1,2,3}\, \sigma_{0i} \right\}, \tag{2}$$

$$\det\left|\overline{\overline{\sigma}}(t) - \boldsymbol{\sigma_O}\overline{\overline{I}}\right| = 0, \tag{3}$$

$$\boldsymbol{P_n}(\phi_n, \theta_n, t) = \overline{\overline{\sigma}}(t)\, \boldsymbol{n}(\phi_n, \theta_n), \tag{4}$$

$$\boldsymbol{\sigma_n}(\phi_n, \theta_n, t) = \boldsymbol{n}^T(\phi_n, \theta_n)\overline{\overline{\sigma}}(t)\, \boldsymbol{n}(\phi_n, \theta_n), \tag{5}$$

$$\boldsymbol{\tau_n}(\phi_n, \theta_n, t) = \boldsymbol{u}^T(\phi_n, \theta_n)\overline{\overline{\sigma}}(t)\boldsymbol{u}(\phi_n, \theta_n) + \boldsymbol{v}^T(\phi_n, \theta_n)\overline{\overline{\sigma}}(t)\,\boldsymbol{v}(\phi_n, \theta_n), \tag{6}$$

$$\boldsymbol{n}(\phi_n, \theta_n) = \begin{bmatrix} \cos\phi_n \sin\theta_n \\ \sin\phi_n \sin\theta_n \\ \cos\theta_n \end{bmatrix}; \boldsymbol{u}(\phi_n, \theta_n) = \begin{bmatrix} -\sin\theta_n \\ \cos\phi_n \\ 0 \end{bmatrix}; \boldsymbol{v}(\phi_n, \theta_n) = \begin{bmatrix} -\cos\phi_n \cos\theta_n \\ -\sin\phi_n \cos\theta_n \\ \sin\theta_n \end{bmatrix}. \tag{7}$$

For periodic stresses, $\boldsymbol{P_n}$ describes a closed curve in the space and, therefore, $\boldsymbol{\tau_n}$ describes a closed curve in the plane. This curve is called as Γ_n (Fig. 1(b)). With respect to $\boldsymbol{\sigma_n}$, along the period T, it assumes different values from a minimum $\sigma_{n,min}$ to a maximum $\sigma_{n,max}$ (Fig. 1(b)). Therefore, it is possible to define the value of the alternating stress (acting on the plane having normal $\boldsymbol{n}$) $\sigma_{n,a}$ according to eqn (8).

$$\sigma_{n,a} = \max_T\{\boldsymbol{\sigma_n}(t)\} - \min_T\{\boldsymbol{\sigma_n}(t)\} = \sigma_{n,max} - \sigma_{n,min}. \tag{8}$$

The curve Γ_n is representative of the tangential stresses acting on the studied plane during the entire loading cycle. To translate Γ_n into a value of alternate tangential stress (exerting on the plane with normal $\boldsymbol{n}$) $\tau_{n,a}$ several methods can be found in the literature. The most diffused method is the Minimum Circumscribed Circle (MCC) (eqn (9)) [34], i.e. $\tau_{n,a}$ is calculated as the radius of the smallest circle that can entirely contain the curve Γ_n (Fig. 2).

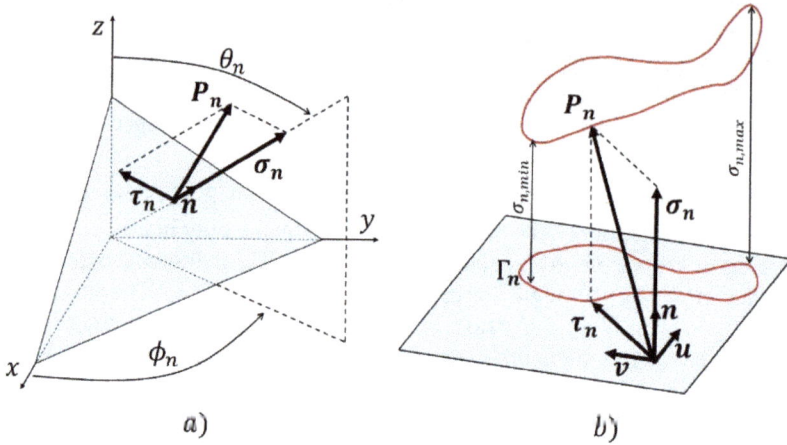

Figure 1: (a) Components of $P_n(\phi_n, \theta_n, t)$ on the plane $n(\phi_n, \theta_n)$; and (b) Definition of the curve Γ_n.

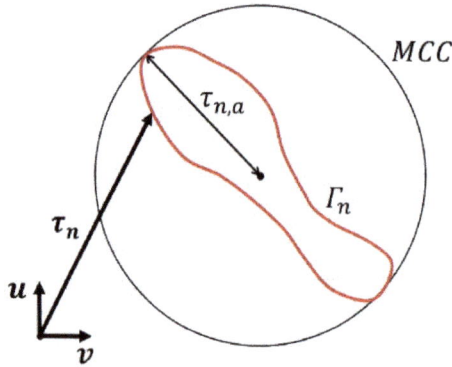

Figure 2: Minimum Circumscribed Circle (MCC) method.

$$\tau_{n,a} = \underset{T}{MCC}\{\boldsymbol{\tau_n}(t)\}. \tag{9}$$

For each plane (having normal $\boldsymbol{n}$), that can be defined by varying the parameters (ϕ_n, θ_n), it is possible to calculate the relevant stress parameters, i.e. $\tau_{n,a}$, $\sigma_{n,max}$, and $\sigma_{n,a}$. More specifically, for the critical plane, the corresponding spherical coordinates and the related stresses will be labelled with the subscript c, i.e. ϕ_c, θ_c, $\tau_{c,a}$, $\sigma_{c,max}$, and $\sigma_{c,a}$.

The damage parameter of each fatigue criterion based on the critical plane can be represented as in eqn (10). In eqn (10), the parameter k is a constant related to the material properties and S is a variable related to the normal stresses exerting on the critical plane. Both of them varies according to the fatigue criterion in question. A summary of how the k and S parameters are defined based on the different fatigue criteria can be found in eqns (11)–(15).

In particular, it is possible to notice that for the calculation of the parameter k according to the McDiarmid criterion, the stress-rupture σ_R is involved. For the other fatigue criteria, k can be calculated through the material fatigue limit at symmetrical alternating bending loading (σ_f), and the material fatigue limit at symmetrical alternating torsional loading (τ_f). With respect to the variable S, the Papadopoulos criterion considers the maximum octahedral stress $\sigma_{h,max}$ while, the others, consider stresses related to the critical plane, i.e. $\sigma_{c,max}$ in Finley and McDiarmid, $\sigma_{c,a}$ in Matake, and $\sigma_{c,max}/\tau_{c,a}$ in Susmel et al.

$$DP = \tau_{c,a} + kS, \tag{10}$$

$$DP_{Findley} = \tau_{c,a} + \frac{2r_{\tau/\sigma}-1}{2\left(\sqrt{r_{\tau/\sigma}-r_{\tau/\sigma}^2}\right)}\sigma_{c,max}, \tag{11}$$

$$DP_{Matake} = \tau_{c,a} + \left(2r_{\tau/\sigma} - 1\right)\sigma_{c,a}, \tag{12}$$

$$DP_{Susmel\ et\ al.} = \tau_{c,a} + \left(\tau_f - \frac{\sigma_f}{2}\right)\frac{\sigma_{c,max}}{\tau_{c,a}}, \tag{13}$$

$$DP_{Papadopoulos} = \tau_{c,a} + \left(\frac{3}{2}\left(2r_{\tau/\sigma} - 1\right)\right)\sigma_{h,max}, \tag{14}$$

$$DP_{McDiarmid} = \tau_{c,a} + \frac{\tau_f}{2\sigma_R}\sigma_{c,max}, \tag{15}$$

where

$$r_{\tau/\sigma} = \tau_f/\sigma_f. \tag{16}$$

The determination of the critical plane (ϕ_c, θ_c) differs for the different fatigue criteria. For the Findley criterion, the critical plane is the plane on which the damage parameter assumes its maximum value (eqn (17)) while, for the other fatigue criteria, the critical plane coincides with the plane on which the $\tau_{c,a}$ assumes its maximum value (eqn (18)). Therefore, the application of the Findley criterion could lead to the identification of a critical plane having a different orientation with respect to the critical plane found applying the other fatigue criteria.

$$(\phi_c, \theta_c) \rightarrow \max_{\phi,\theta}\{\tau_{n,a}(\phi,\theta) + kS(\phi,\theta)\}, \tag{17}$$

$$(\phi_c, \theta_c) \rightarrow \max_{\phi,\theta}\{\tau_{n,a}(\phi,\theta)\}. \tag{18}$$

3 MATERIAL AND METHOD

3.1 Presentation of the general approach

The concept behind the proposed approach is to simulate, for the same geometry, the STBF and the RG conditions with applied loads that lead, according to [6], to the same σ_{Flim}. Hence, through the results of the Finite Element (FE) models, it is possible to obtain the stresses histories (in terms of stress tensors $\overline{\overline{\sigma}}(t)$) for all the nodes N in the tooth root fillet

WIT Transactions on Engineering Sciences, Vol 130, © 2021 WIT Press
www.witpress.com, ISSN 1743-3533 (on-line)

region, i.e. each point where fracture could nucleate. Therefore, by analyzing the stress histories through a fatigue criterion based on a critical plane, it is possible to:

1. Individuate the critical plane for each point ($\theta_c \phi_c(N)$);
2. Evaluate the damage parameter in each critical plane (eqn (10) applied on each node); and
3. Identify the critical point in which the damage parameter assumes the maximum value ($\max_{N}\{(\tau_{c,a} + kS)\}$).

This process can be followed for the RG and the STBF simulations. The ratio between the maximum damage parameter emerged in the STBF condition and the one observed for the RG condition corresponds to the f_{korr} as it represents the ratio between the different effects that cause failure for tooth bending failure (eqn (19)). The overall approach can be carried out by applying different fatigue criteria. More detail on the FE simulations and on the implementation of the above-mentioned fatigue criteria can be found in the following sections.

$$f_{korr} = \frac{\max_{N}\{(\tau_{c,a}+kS)\}|_{STBF}}{\max_{N}\{(\tau_{c,a}+kS)\}|_{RG}}. \tag{19}$$

3.2 Finite element analysis

In the present work, the gear geometry presented in [28] has been modelled in both the RG and STBF conditions. Through KISSsoft® a CAD model of the gear (having the parameters listed in Table 1) has been realized. Hence, this model was imported into the open source FE software Salome-Meca/Code_Aster where the RG and STBF conditions have been numerically reproduced.

Table 1: Geometrical parameter of the simulated gear according to [28].

Description	Symbol	Unit	Value
Normal module	m_n	[mm]	4
Normal pressure angle	α_n	[°]	20
Number of teeth	z	[-]	28
Face width	b	[mm]	30
Profile shift coefficient	x	[-]	0
Dedendum coefficient	h_{fP}^*	[-]	1.25
Root radius factor	ρ_{fP}^*	[-]	0.38
Addendum coefficient	h_{aP}^*	[-]	1

In the FE model, symmetries were exploited to reduce the computational effort. In particular, in STBF simulations a quarter of each gear has been meshed while in the RG ones the whole gear profile has been modelled for half of the width. For each model, an extruded mesh was created. The quality of the mesh has been improved in teeth subjected to loads. More specifically, in that region the mesh density was increased and hexahedral elements have been exploited (Fig. 3). Non-linear simulations have been carried out setting 40 time-steps (along the period T) for loading cycle.

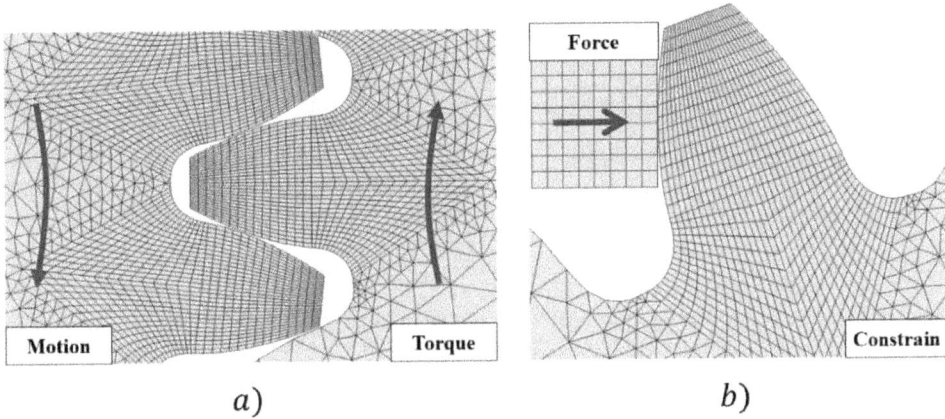

Figure 3: Finite element models of (a) RG; and (b) STBF tests.

The model of the RG and STBF configurations are represented in Fig 1(a) and 1(b) respectively. In RG, the engaging gears were positioned with the appropriate center distance and the axes of rotation were fixed. The motion was assigned to the driving gear and a resistant torque to the driven once. In STBF, the radial symmetry was exploited and a pulsating force with R = 0.1 was applied to the anvil. In particular, the force (in STBF) and the torque (in RG) have been set in order to lead to the same σ_F according to the standard ISO 6336-3 [6]. Typical steels properties have been applied to the components. For each simulation, the $\overline{\overline{\sigma}}(t)$ in the nodes in the tooth root fillet region have been extracted.

3.3 Framework to calculate f_{korr} applying different fatigue criteria

Since the FE simulations were performed by setting the properties of steels in the linear elastic range, the results (in terms of $\overline{\overline{\sigma}}(t)$) can be used to analyze any steel. In this case, three steels were analyzed (i.e. 18NiCrMo5, 42CrMoS4, and 31CrMo12) since these steels were also exploited in the gears tested in [28]. In Table 2, it is possible to see the fatigue limits (according to [35]) and the rupture stresses (according to [28]). These data were elaborated to calculate the parameter k according to the various fatigue criteria through the formulas shown in eqns (11) to (15). Preliminary results are provided in Table 2.

Table 2: Constants related to the material properties.

Material	σ_f	τ_f	σ_R	k according to				
				Findley	Matake	Susmel	Papad.	McDiar.
18NiCrMo5	660.7	342.7	1467	0.037	0.037	12.350	0.056	0.117
42CrMoS4	525.7	336.3	1160	0.291	0.279	73.450	0.419	0.145
31CrMo12	628.3	366.6	987	0.169	0.167	52.450	0.250	0.186

The f_{korr} (according with the different fatigue criteria) have been calculated following the workflow presented in Fig. 4.

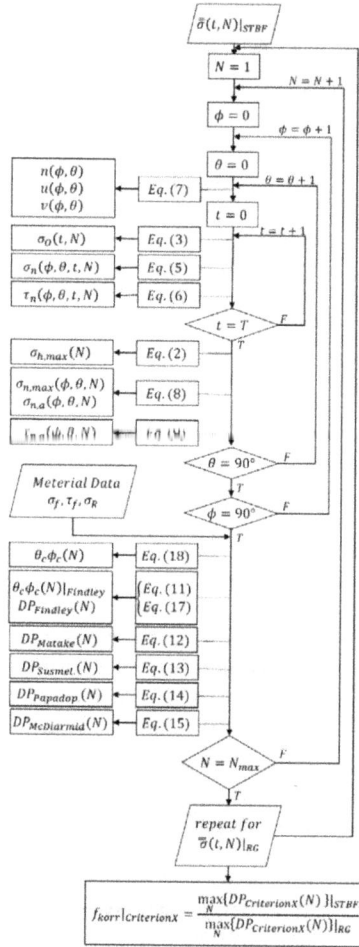

Figure 4: Framework to calculate f_{korr} according to the different fatigue criteria.

For each simulation, the workflow is structured with four FOR loops. The innermost one analyses data along the time from time 0 to the end of cycle T for each time step recorded. The FOR loops on θ and ϕ aim to discretize the space of potential critical planes, in this case with a resolution of one degree. The FOR loop on the nodes (in this case 16 nodes within the tooth root fillet i.e. $N_{max} = 16$), aims to identify the most stressed node according to the various fatigue criteria. Eventually, the f_{korr} (calculated based on the implementation of a specific fatigue criterion) have been calculated through the ratio between the maximum value of the damage parameter recorded in STBF nodes and the ones emerged in RG nodes.

4 RESULTS AND DISCUSSION

The results of the implementation of the proposed approach, in terms of f_{korr} at varying fatigue criterion and material, are summarized in Table 3. The results underline that, for the studied geometry, the application of the Susmel et al. criterion generally leads to higher values of f_{korr} (ranging from 0.87 to 0.99) while the application of the Matake criterion leads to lower values of f_{korr} (ranging from 0.77 to 0.79). With respect to the studied materials, the Matake and the McDiarmid show the lower variability of f_{korr}. On the other hand, the Findley and the Susmel et al. criterion show the greater variability. Therefore, the application of different fatigue criteria on the same gear (in terms of geometry and material) lead to different results. In particular, the maximum variability occurs for 42CrMoS4 where f_{korr} ranges from 0.77 to 0.99.

Table 3: f_{korr} calculated through different combinations of materials and fatigue criteria.

Material	f_{korr} according to				
	Findley	Matake	Susmel et al.	Papadopoulos	McDiarmid
18NiCrMo5	0.81	0.79	0.87	0.81	0.83
42CrMoS4	0.91	0.77	0.99	0.86	0.83
31CrMo12	0.86	0.78	0.97	0.84	0.84

Further considerations can be made based on the results shown in Fig. 5. In Fig. 5, the critical plane passing through the critical node has been reported for RG and STBF conditions. In the figure, the fifteen combination of material and fatigue criterion studied are reported. In all these combinations, the angle θ_c has resulted 90°. Therefore, it has been possible to report the solutions in a two-dimensional representation.

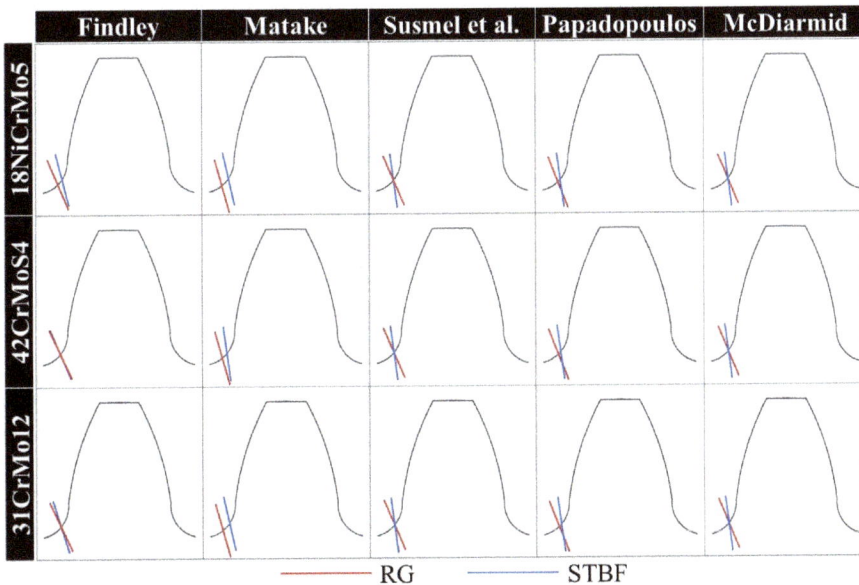

Figure 5:   Comparison between the critical planes emerged for RG and STBF for different material–criterion combinations.

It is worth noting that the critical node in the STBF condition can be different from the critical node in the RG condition. If this occurs (i.e. Findley-18NiCrMo5, Matake and Papadopoulos for all materials), the critical node related to the STBF condition is located closer to the meshing area than the critical node related to the RG condition. Moreover, it is possible to notice that the Matake criterion lead to the largest discrepancy, in terms of critical node location, between the RG and STBF conditions. Furthermore, the results clearly highlight that, in the RG condition, the angle between the critical plane and the tooth axis is always greater than in STBF condition.

5 CONCLUSIONS

In conclusion, this paper describes an approach to calculate a coefficient (f_{korr}) that enables the translation of STBF test results into usable data for gear design. This approach proposes to analyse the results of FE simulations (which aim to represent STBF and RG conditions leading to the same σ_F according to ISO 6336) with different fatigue criteria based on the critical plane. In particular, the implementation of these fatigue criteria requires data on specific material properties (i.e. τ_f, σ_f, and σ_R) that can be obtained from standard tests or from literature. Moreover, through the implementation of this approach, it is possible to obtain the position of the most critical node and the orientation of its critical plane for each simulated condition.

This approach was applied to a specific gear geometry exploited in previous studies by the authors' research group. In addition, the application of different fatigue criteria (i.e. Findley, Matake, McDiarmid, Papadopoulos, and Susmel et al.) as well as different materials (i.e. 18NiCrMo5, 42CrMoS4, and 31CrMo12) have been investigated in this paper.

Results show that the fatigue criterion exploited has a considerable impact on the value of f_{korr}. Therefore, for the purpose of this paper, the fatigue criteria are not equivalent to each other. Future studies should aim to investigate which fatigue criterion is the most appropriate.

REFERENCES

[1]   Vullo, V., *Gears*, Springer International Publishing, 2020.
[2]   Radzevich, S.P. & Dudley, D.W., *Handbook of Practical Gear Design*, CRC Press, 1994.
[3]   Fernandes, P.J.L., Tooth bending fatigue failures in gears. *Engineering Failure Analysis*, **3**(3), pp. 219–225, 1996. DOI: 10.1016/1350-6307(96)00008-8.
[4]   Pantazopoulos, G.A., Bending fatigue failure of a helical pinion bevel gear. *Journal of Failure Analysis and Prevention*, **15**(2), pp. 219–226, 2015.
      DOI: 10.1007/s11668-015-9947-2.
[5]   Bretl, N., Schurer, S., Tobie, T., Stahl, K. & Höhn, B.R., Investigations on tooth root bending strength of case hardened gears in the range of high cycle fatigue. *American Gear Manufacturers Association Fall Technical Meeting*, pp. 103–118, 2013.
[6]   Hong, I.J., Kahraman, A. & Anderson, N., A rotating gear test methodology for evaluation of high-cycle tooth bending fatigue lives under fully reversed and fully released loading conditions. *International Journal of Fatigue*, **133**, p. 105432, 2020.
      DOI: 10.1016/j.ijfatigue.2019.105432.
[7]   ISO 6336-3:2006, *Calculation of Load Capacity of Spur and Helical Gears, Part 3: Calculation of Tooth Bending Strength*, Standard: Geneva, CH, 2006.
[8]   ANSI/AGMA 2001-D04, *Fundamental Rating Factors and Calculation Methods for Involute Spur and Helical Gear Teeth*, American Gear Manufacturers Association: Alexandria, 2004.
[9]   Rao, S.B. & McPherson, D.R., Experimental characterization of bending fatigue strength in gear teeth. *Gear Technology*, **20**(1), pp. 25–32, 2003.

[10] Benedetti, M., Fontanari, V., Höhn, B.R., Oster, P. & Tobie, T., Influence of shot peening on bending tooth fatigue limit of case hardened gears. *International Journal of Fatigue*, **24**(11), pp. 1127–1136, 2002. DOI: 10.1016/S0142-1123(02)00034-8.

[11] McPherson, D.R. & Rao, S.B., Methodology for translating single-tooth bending fatigue data to be comparable to running gear data. *Gear Technology*, pp. 42–51, 2008.

[12] Dobler, D.I.A., Hergesell, I.M. & Stahl, I.K., Increased tooth bending strength and pitting load capacity of fine-module gears. *Gear Technology*, **33**(7), pp. 48–53, 2016.

[13] Concli, F., Tooth root bending strength of gears: Dimensional effect for small gears having a module below 5 mm. *Applied Science*, **11**, p. 2416, 2021. DOI: 10.3390/app11052416.

[14] Gorla, C., Conrado, E., Rosa, F. & Concli, F., Contact and bending fatigue behaviour of austempered ductile iron gears. *Proceedings of the Institution of Mechanical Engineers, Part C: Journal of Mechanical Engineering Science*, **232**(6), pp. 998–1008, 2018. DOI: 10.1177/0954406217695846.

[15] McPherson, D.R. & Rao, S.B., *Mechanical Testing of Gears*, ASM International: Materials Park, OH, pp. 861–872, 2000.

[16] Rettig, H., Ermittlung von Zahnfußfestigkeitskennwerten auf Verspannungsprüfständen und Pulsatoren-Vergleich der Prüfverfahren und der gewonnenen Kennwerte. *Antriebstechnik*, **26**, pp. 51–55, 1987.

[17] Stahl, K., Lebensdauer statistik: Abschlussbericht, forschungsvorhaben nr. 304. Technical Report, 580, 1999.

[18] Concli, F., Fraccaroli, L. & Maccioni, L., Gear root bending strength: A new multiaxial approach to translate the results of single tooth bending fatigue tests to meshing gears. *Metals,* **11**(6), p. 863. DOI: 10.3390/met11060863.

[19] Concli, F., Austempered Ductile Iron (ADI) for gears: Contact and bending fatigue behavior. *Procedia Structural Integrity*, **8**, pp. 14–23, 2018. DOI: 10.1016/j.prostr.2017.12.003.

[20] Bonaiti, L., Concli, F., Gorla, C. & Rosa, F., Bending fatigue behaviour of 17-4 PH gears produced via selective laser melting. *Procedia Structural Integrity*, **24**, pp. 764–774, 2019. DOI: 10.1016/j.prostr.2020.02.068.

[21] Gasparini, G., Mariani, U., Gorla, C., Filippini, M. & Rosa, F., Bending 367 fatigue tests of helicopter case carburized gears: Influence of material, design 368 and manufacturing parameters. *American Gear Manufacturers Association 369 (AGMA) Fall Technical Meeting*, pp. 131–142, 2008.

[22] Gorla, C., Rosa, F., Concli, F. & Albertini, H., Bending fatigue strength of innovative gear materials for wind turbines gearboxes: Effect of surface coatings. *ASME International Mechanical Engineering Congress and Exposition*, vol. 45233, American Society of Mechanical Engineers, pp. 3141–3147, 2012. DOI: 10.1115/IMECE2012-86513.

[23] Gorla, C., Rosa, F., Conrado, E. & Concli, F., Bending fatigue strength of case carburized and nitrided gear steels for aeronautical applications. *International Journal of Applied Engineering Research*, **12**(21), pp. 11306–11322, 2017.

[24] Rao, S.B., Schwanger, V., McPherson, D.R. & Rudd, C., Measurement and validation of dynamic bending stresses in spur gear teeth. *International Design Engineering Technical Conferences and Computers and Information in Engineering Conference*, vol. 4742, pp. 755–764, 2005. DOI: 10.1115/DETC2005-84419.

[25] Wagner, M., Isaacson, A., Knox, K. & Hylton, T., Single tooth bending fatigue testing at any R ratio. *2020 AGMA/ABMA Annual Meeting*, AGMA American Gear Manufacturers Association, 2020.

[26] Bonaiti, L., Bayoumi, A.B.M., Concli, F., Rosa, F. & Gorla, C., Gear root bending strength: A comparison between single tooth bending fatigue tests and meshing gears. *Journal of Mechanical Design*, pp. 1–17, 2021. DOI: 10.1115/1.4050560.

[27] Crossland, B., Effect of large hydrostatic pressures on the torsional fatigue strength of an alloy steel. *Proceedings of the International Conference on Fatigue of Metals*, vol. 138, Institution of Mechanical Engineers: London, pp. 12–12, 1956.

[28] Conrado, E., Gorla, C., Davoli, P. & Boniardi, M., A comparison of bending fatigue strength of carburized and nitrided gears for industrial applications. *Engineering Failure Analysis*, **78**, pp. 41–54, 2017. DOI: 10.1016/j.engfailanal.2017.03.006.

[29] Findley, W.N., A theory for the effect of mean stress on fatigue of metals under combined torsion and axial load or bending. *Journal of Engineering for Industry*, **81**(4), pp. 301–305, 1959. DOI: 10.1115/1.4008327.

[30] Matake, T., An explanation on fatigue limit under combined stress. *Bulletin of JSME*, **20**(141), pp. 257–263, 1977. DOI: 10.1299/jsme1958.20.257.

[31] McDiarmid, D.L., Fatigue under out-of-phase biaxial stresses of different frequencies. *Multiaxial Fatigue*, ASTM International, 1985. DOI: 10.1520/STP36245S.

[32] Papadopoulos, I.V., A high-cycle fatigue criterion applied in biaxial and triaxial out-of-phase stress conditions. *Fatigue & Fracture of Engineering Materials & Structures*, **18**(1), pp. 79–91, 1995. DOI: 10.1111/j.1460-2695.1995.tb00143.x.

[33] Susmel, L., Tovo, R. & Lazzarin, P., The mean stress effect on the high-cycle fatigue strength from a multiaxial fatigue point of view. *International Journal of Fatigue*, **27**(8), pp. 928–943, 2005. DOI: 10.1016/j.ijfatigue.2004.11.012.

[34] Papadopoulos, I.V., Critical plane approaches in high-cycle fatigue: On the definition of the amplitude and mean value of the shear stress acting on the critical plane. *Fatigue & Fracture of Engineering Materials & Structures*, **21**(3), pp. 269–285, 1998. DOI: 10.1046/j.1460-2695.1998.00459.x.

[35] Susmel, L., On the overall accuracy of the modified Wöhler curve method in estimating high-cycle multiaxial fatigue strength. *Frattura ed Integrita Strutturale*, **5**(16), pp. 5–17, 2011. DOI: 10.3221/IGF-ESIS.16.01.

CHARACTERIZATION OF THE CRACK PROPAGATION IN A MICROSTRUCTURALLY RANDOM MATERIAL

MIGUEL A. RODRIGUEZ MARQUEZ[1], CARLOS A. MORA SANTOS[1], HELVIO R. MOLLINEDO[2],
JUDITH DIAZ DOMINGUEZ[1] & JORGE BEDOLLA HERNÁNDEZ[1]
[1]Tecnológico Nacional De México, Instituto Tecnológico de Apizaco, Mexico
[2]Instituto Politécnico Nacional, Unidad Interdisciplinaria en Ingeniería y Tecnologías Avanzadas, Mexico

ABSTRACT

In the study of mechanical properties of materials the microstructure of a material is usually subjected to some kind of homogenization; however, there are materials in which the microstructural disorder must be considered. This disorder manifests itself in the fracture resistance of materials. Some empirical experimental studies and various types of models (based on variations in mass per unit area) have been made to relate the effect of the disorder during crack propagation with the macroscopic resistance of the material, but the absolute-density/mass projections have not been a good descriptor to extrapolate the behavior of the material between its microstructure and the macroscale since it is difficult to determine the porosity and the net trajectory of the fibers. The physical phenomenon of the instability of the crack propagation of interest in the present work occurs on a meso-scale, where the microstructure of the materials can be characterized only statistically and has been established as the range in which the bridge can exist between the micro and macro behavior of this kind of materials. By the Digital Image Correlation Technique the crack propagation is followed based on the displacements produced locally by the arrangement of the fibers in front of the crack tip of paper, as a material model. At the beginning of the load process is observed a smooth trace in the peak local deformation corresponding with the elastic part of the stress-curve; after, when the stress-curve starts to deflect, the peak local-deformation trace change in its slope and it becomes intermittent, this behavior is attributed to the local conditions of material. Finally, it observed that the local deformation is a good descriptor for the crack extension.
Keywords: *local deformation, crack path, inhomogeneity.*

1 INTRODUCTION

Fracture has been one of the phenomena that has attracted the most attention in the scientific and engineering community of materials. As a failure condition, it limits the function for which a material was developed, since it can be understood microscopically, as interatomic separation, while at the macro level, as rupture in two or more parts of the material.

Even though the study of the effect of fracture on the strength of materials dates from the pioneering works of DaVinci, it is only from the 19th century, when it has been based on two principles based on the microstructural formation of materials: from the Classical *Theory of Continuous Media* and recently the *Size Effect*. In the first case, homogeneity and continuity are assumed and, the microstructural effect of the material is lost; while in the second, when the size effect is taken into account, a microstructural dependence of the material must be considered. This limitation has once again drawn attention in its application to heterogeneous materials. That is, in heterogeneous materials of complex structure such as paper, it is known that structural inhomogeneity (called formation) has an influence on the mechanical properties of strength and fracture. The structural disorder manifests itself in the variation of the mass density, the elastic coefficient and the local stresses of the material. These variations are presented in a short range order, but they could not be correlated on larger scales (long range).

Due to the heterogeneity caused by the complex fiber arrangement which make up the paper, this material has been a good descriptor of the effect of disorder and it has been found

WIT Transactions on Engineering Sciences, Vol 130, © 2021 WIT Press
www.witpress.com, ISSN 1743-3533 (on-line)
doi:10.2495/CMEM210111

in many applications as a model material in studies of disordered phenomena such as fracture [1]–[3]. In order to relate the fibers disorder during the crack propagation with the macroscopical strength of the material have been carried out some experimental studies and generated several kind of empirical models to predict the fracture mainly based on the variations *mass/unit area*, but the *absolute density/mass* projections have not been a good descriptor to extrapolate the behavior of the material between its microstructure and the macroscale since it is difficult determining the porosity and the net trajectory of the fibers.

During the fracture of the paper, the load causes an initial sliding between fibers and interfiber bonding (considered as two different phases due to the absence of a matrix) until the catastrophic failure of the material is finally reached with the breakdown of both phases [4]. These mechanisms are factor of energy consume during the crack propagation and related with the area of the microscopic damage and of the local deformation [5]–[8].

In this work the crack propagation in paper is studied. In the present work is considered that during the initiation of crack growth, the inhomogeneity, as a precursor to the imminent propagation of cracks, can be determined from local displacements on a meso-scale (scale between millimeters and centimeters) around the crack tip and within which the local variations in the field of displacements correlate with the overall strength properties of the sample.

2 EXPERIMENT

In order to relate the micro and macro behavior of fracture, a synchronization between a universal tensile testing machine and a high resolution/rate camera was done, as shown in Fig. 1.

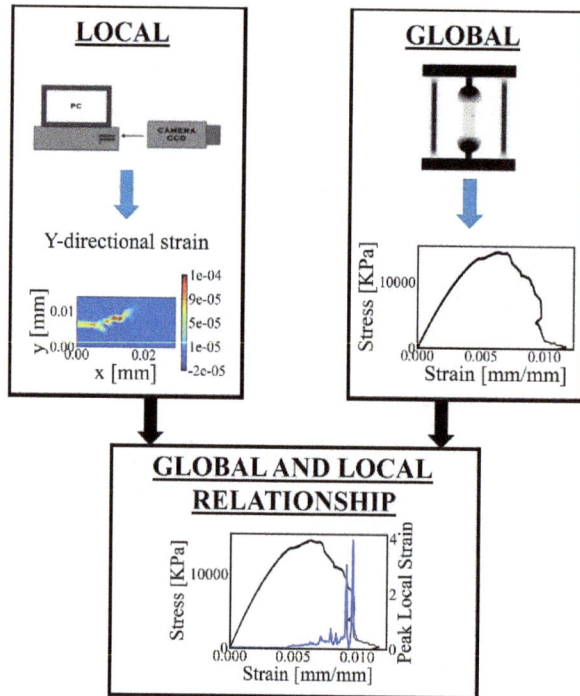

Figure 1: Experimental procedure.

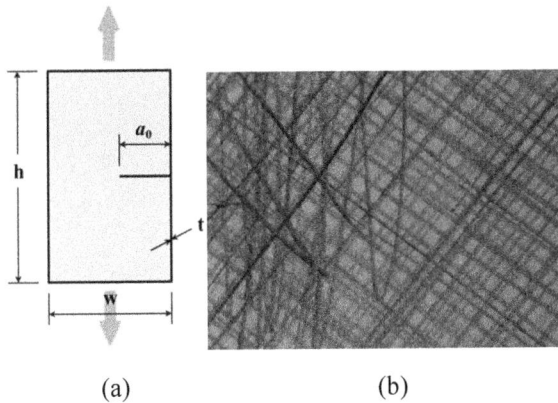

Figure 2:   Specimen configuration. (a) Single edge notch testing; and (b) Scratched sample for the images acquiring.

Here, the time is considered as the common term for the tests. By the universal tensile testing machine which run at a constant displacement rate, the stress–strain curve is obtained and from which the global fracture behavior of the material is studied: while whit the high resolution camera n-images are obtained and from which the local deformation maps are calculated with correlation numerical techniques.

Single Edge Notch Tension (SENT) specimens of commercial paper, with constant width of w = 50 mm, total height h = 100 mm, and thickness t = 10.07 μm with a variable initial notch (a_0 = 1, 5, 20 and 40 mm) were considered, as shown in Fig. 2. In order to get the randomness in the grey colour for the images the specimens were prepared by scratching. The testing were done by an axial load until to reach the total rupture of the specimens in an electrodynamical universal tensile testing machine of 1 kN at a 0.01 mm/s under similar conditions of Room Temperature of 22.5 ± 0.11°C and Relative Humidity of 27.94 ± 0.2%. The images, with a resulting size of 1027 × 767 pixel (29 mm × 21.7 mm), where 1 pixel = 28.34 μm, were got only during the lapse that each testing lasts and at a 10 images/s rate (the region of study final was 5.7 × 7.1 mm). Therefore, with the considered experimental set up results that 10 images = 0.01 mm of constant displacement between clamps. The image correlations were based on the elastic image registration procedure [9] by the sequential comparison between a reference image and the deformed image.

3 RESULTS

3.1 Macroscale behavior

By the stress–strain diagrams the macroscopic behavior is observed for different sizes of initial crack in Fig. 3, paper develops an elastoplastic constitutive model (Fig. 3(a)). The tensile strength is dependent of the length of the initial notch (Fig. 3(b)); that is, in an increase of the initial notch there are a reduction in the ligament area which increasing the stress intensity to reach the strength of the material. That is why this parameter is in inconvenient to use as a design parameter in heterogeneous materials.

The stiffness is the capacity which have a system to withstand loads without excessive deformations. Based on the stress–strain curves it is observed two behaviors: the stiffness is constant and independent for the relation $a_0/w \leq 0.1$, while for $a_0/w > 0.1$ a linear dependence of the elastic modulus E with the initial notch is observed; that is, $E = E(a_0/w)$ as shown in Fig. 3(c).

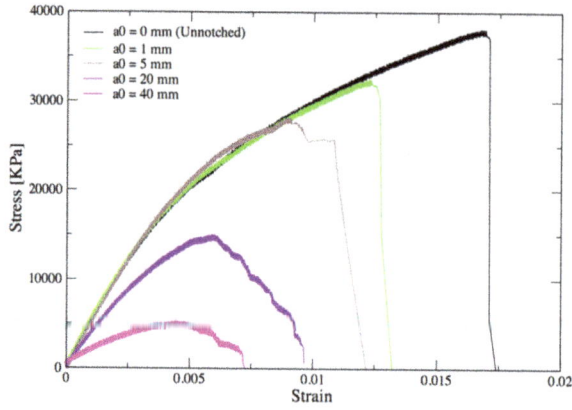

(a)

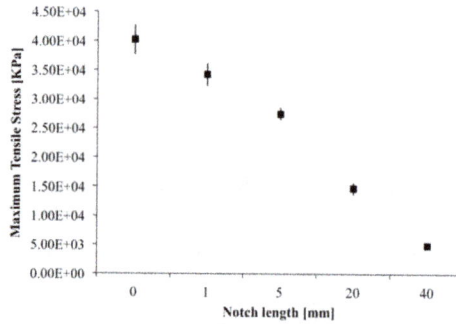

(b)

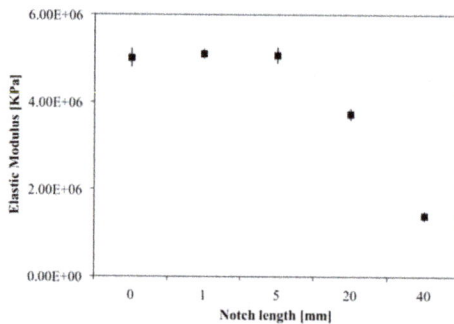

(c)

Figure 3:   Macroscopic fracture behavior of the material for several initial notch size. (a) Typical stress–strain curves; (b) Tensile strength; and (c) Elastic modulus.

3.2 Fracture process in paper

The microstructural disorder makes the study of the fracture in the paper complex, for example one of the main difficulties in the characterization of cracks is the specification of the instant in which the localized crack get started to grow [4]; also, it causes that the path of the propagation be random which difficult to follow.

The fracture micromechanisms in this type of materials makes an intensification in the local strain in front of the crack [6]–[8] and it can be measured and followed each instant for all fracture process through the Digital Image Correlation Method (DIC). In this work, the *Peak Local Deformation* (PLD) was defined as the parameter to characterize the crack behavior and its propagation at a meso-scale (Fig. 4(a)). However, due to the existence of small intrinsic deformations distributed over all the specimen and produce noise in the measurements it was considered only the fractile 0.9 of deformation in order to only ensure the use of the localized deformations in front of the crack tip. Fig. 4(b) shows that the PLD is inversely related with the macroscopic behavior of the material; that is in an increasing of the intensity of PLD the stress–strain curve deflects.

(a)

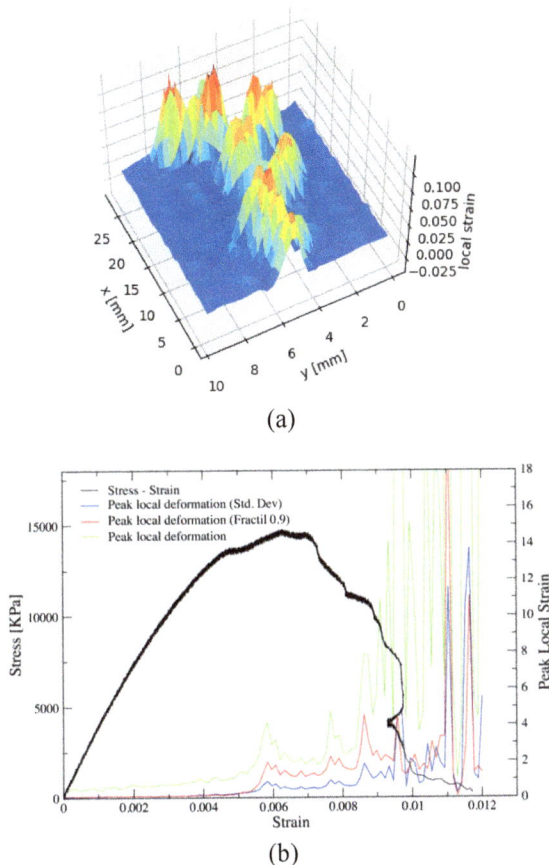

(b)

Figure 4:   Behavior of a crack at meso-scale. (a) Local strain of a developed crack; and (b) Relationship between the Peal Local Deformation (PLD) and the stress–strain curve.

In the resulting deformation maps is observed that once the crack initiate to grow there is a sudden increase in the maximum strain and a random behavior in the path of the crack advancing due to microstructural disorder of the material (Fig. 5).

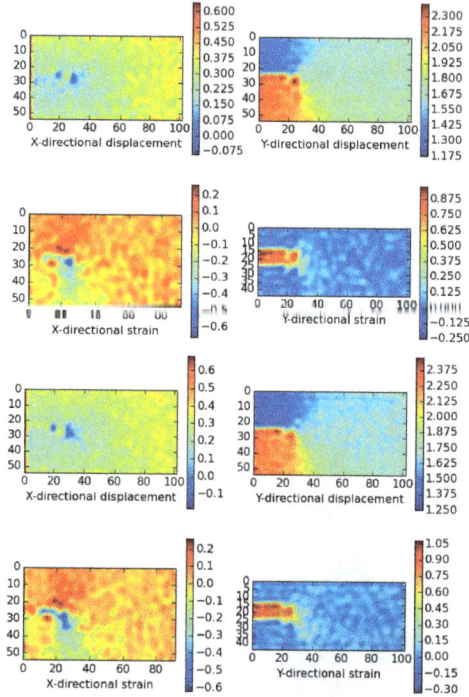

Figure 5: Displacement and strain fields at the instant in the initiation of the crack to grow.

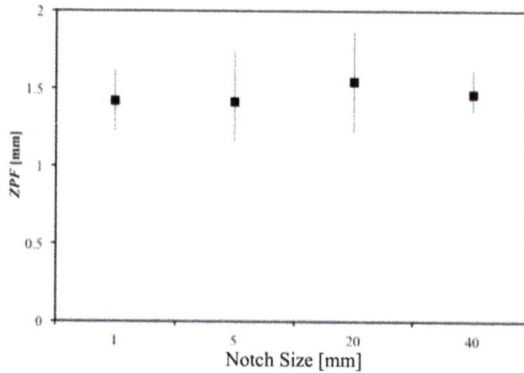

Figure 6: Fracture process zone behavior for several initial sizes of notch a_0.

The fracture in materials with complex microstructure is basically defined by a localized crack and a *Fracture Process Zone* (FPZ). This is a diffuse area which is characterized by its width, length and depends on the load, and microstructural disorder of the material [10], [11]. With the experimental process used here the FPZ was measured. The formation of the FPZ is when the stress–strain curve lost its first linear slope and developed until the beginning of stable crack propagation. By the local strain intensity this phenomenon is shown as a sudden peak (in Fig. 4 the FPZ begins approximately at 0.004). As shown in Fig. 6 there is not dependence of the FPZ on the initial size of the notch a_0, this corroborates the importance of its consideration as a material property.

3.3 Relationship between the meso and macro scales

By setting the peak local deformation around the tip of the crack as a monitoring parameter, both the length and the crack path can be determined, this is carried out by monitoring the specific position of this parameter. It should be mentioned that in order to carry out the measurement, the condition that the specific position must have positive or zero increments on the x axis must be met. Therefore, comparing the resulting crack length with the stress–strain diagram, it is possible to observe such advancing and its effect on the strength of the material during the entire loading process (Fig. 7), an effect that was shown through the intensity of deformation.

The crack onset a_C has been a fracture parameter difficult to get, it has been done under direct observation or with microscopy, however, it has become difficult since the ZPF manifests itself as a diffuse rather than localized zone and the speed with it spreads is rapid. Fig. 8 shows that the instant can be localizable and measurable (based on the peak local deformation, in Fig. 4, this point is approximately at 0.0053 in the strain). In all figures a direct relationship between the stress–strain curve and the crack growth is observed. Once the crack grows, the sudden change in both the slope and the intermittence of the crack length curve is observed, which means that the crack begins to develop and the intermittency is due to disorder of the material.

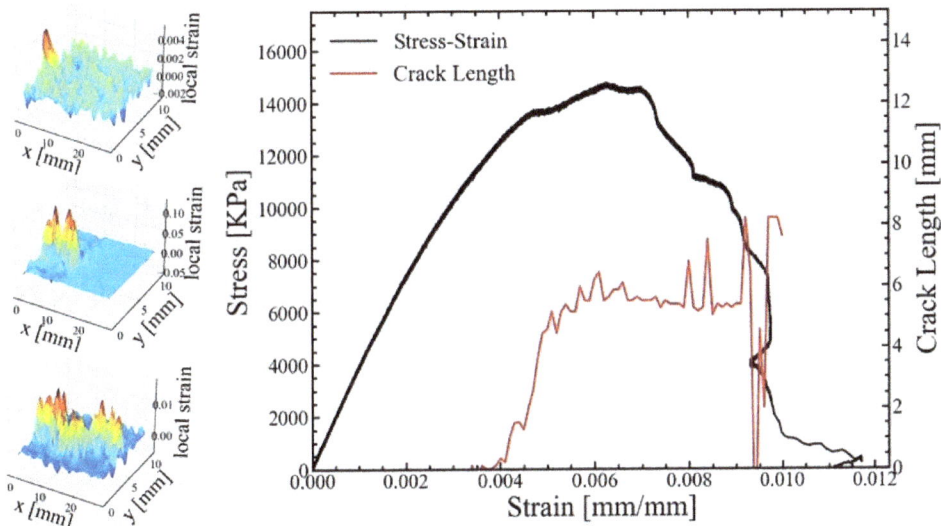

Figure 7: Typical crack length behavior during the fracture process for paper.

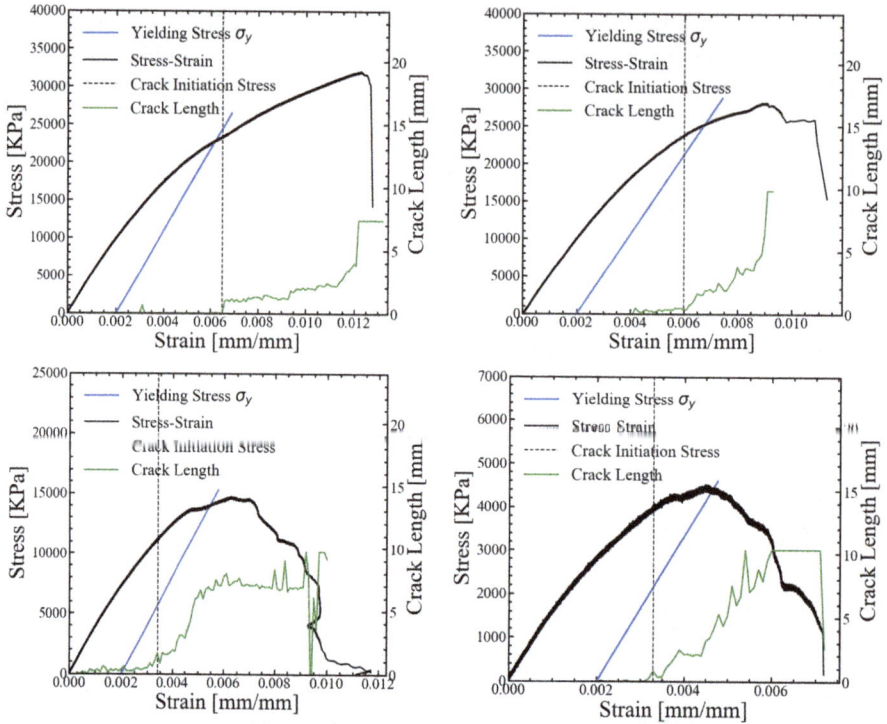

Figure 8: Typical stress–strain/crack length relationships for several initial notch a_0. Vertical line specifies the instant of the crack onset.

For an initial defect of $a_0 = 1$ mm, the crack length necessary to reach the critical rupture, which is catastrophic, is of 3 to 5 mm and in the stress–strain diagram this is imperceptible since the tensile stress is approximately equal to the rupture stress σ_R; while, for the initial notch $a_0 = 40$ mm a crack length of 8 to 12 mm approximately is required, in the stress–strain curve it is observed as a deflected curve. This behavior is due to when the initial notch a_0 is longer the effective area of the specimens is reduced such that the microstructural disorder of the material takes more importance.

For the initial notch range considered the crack onset stress is approximately equal or less than the yielding stress ($\sigma_y = 0.2\%$), this imply that in this kind of materials with crack like defect the yielding stress cannot be used as a design parameter, similar results have been found in other types of paper [4]. When $a_0/w = 0.1$ the crack onset stress is approximately equal to the yielding stress and for a $a_0/w > 0.1$ the crack initiation stress is less than it.

4 CONCLUSIONS

Intrinsic disorder is manifested in the fracture behavior of materials with complex microstructure. When establishing the displacements as descriptors for the characterization of the fracture, it was assumed that the local variations of inhomogeneity are uniform. The DIC technique made it possible to directly evaluate the intensity of the maximum local deformation (PLD) developed around the tip of the crack. Here it was observed that this value is strongly related to the resistance of the material; that is, the increase in PDL is inversely proportional to the resistance of the material and through it, was possible to carry out the monitoring of the crack path until the rupture of material. In all the cases considered here, it was observed that the onset crack stress is equal to or less than the yield stress, depending on the size of the initial defect, which implies that in this type of materials when has a defect, the yield stress cannot be considered as a design parameter. Therefore, based on the above, it is concluded that the displacements are a good descriptor for the characterization of fracture in disordered materials since they do not depend on the mass/unit area relationship.

REFERENCES

[1]   Alava, M. & Niskanen, K., The physics of paper. *Reports on Progress in Physics*, **69**(3), p. 669, 2006.

[2]   Alava, M.J., Nukala, P.K. & Zapperi, S., Statistical models of fracture. *Advances in Physics*, **55**(3–4), pp. 349–476, 2006.

[3]   Balankin, A.S., Susarrey, O., Santos, C.A.M., Patino, J., Yoguez, A. & García, E.I., Stress concentration and size effect in fracture of notched heterogeneous material. *Physical Review E*, **83**(1), p. 015101, 2011.

[4]   Mora Santos, C.A., Susarrey Huerta, O., Flores Lara, V., Bedolla Hernández, J. & Mendoza Núñez, M.A., Failure stress in notched paper sheets. *Key Engineering Material*, **569**, pp. 417–424, 2013.

[5]   Korteoja, M.J., Lukkarinen, A., Kaski, K., Gunderson, D.E., Dahlke, J.L. & Niskanen, K.J., Local strain fields in paper. *Tappi Journal*, **79**(4), pp. 217–222, 1996.

[6]   Seth, R.S., Robertson, A., Mai, X.-W. & Hoffman, J.D., Plane stress fracture toughness of paper. *Tappi Journal*, **76**(2), pp. 109–116, 1993.

[7]   Kettunen, H. & Niskanen, K., Microscopic damage in paper. *Journal of Pulp and Paper Science*, **26**(1), pp. 35–40, 2000.

[8]   Tanaka, A. & Yamauchi, T., Deformation and fracture of paper during the in-plane fracture toughness testing – Examination of the essential work of fracture method. *Journal of Materials Science*, **35**, pp. 1827–1833, 2000.

[9]   Kybic, J. & Unser, M., Fast parametric elastic image registration. *IEEE Transactions on Image Processing*, **12**, pp. 1427–1442, 2003.

[10] Alava, M.J., Nukala, P.K.V.V. & Zapperi, S., Role of disorder in the size scaling of material strength. *Physical Review Letters*, **100**, p. 055502, 2008.

[11] Santucci, S., Cortet, P.P., Deschanel, S., Vanel, L. & Ciliberto, S., Subcritical crack growth in fibrous materials. *Europhysics Letters*, **74**, pp. 595–601, 2006.

SPATIAL CORRELATION OF THE BACKSCATTERED ULTRASONIC GRAIN NOISE IN THE ULTRASONIC INSPECTION OF THE FORGING TITANIUM ALLOY

THEODOR TRANCĂ[1] & IULIANA RADU[2]
[1]DIAC SERVICII SRL, Romania
[2]Zirom SA, Romania

ABSTRACT

Ultrasonic inspection is a routine Non Destructive Examination (NDE) method adopted by the aircraft engine manufacturers. However, the detection of smaller defects in such materials is made difficult by the complicated ultrasound-microstructure interactions. One of the adverse influences of the interactions is the high backscattered grain noise level accompanying the ultrasonic inspections of some titanium alloys. The high grain noise deteriorates the Signal to Noise ratio (S/N) of pulse/echo inspections and consequently may lead to the missing detection of an existing flaw. Ultrasonic signal fluctuations have direct impact on flaw detection, flaw characterization and the estimation of the Probability of Detection (POD). The total backscattering is controlled by grain morphology, grain orientation and elastic anisotropy, which may vary throughout the microstructure. Thus any Thermo-Mechanical Processing (TMP) leading to the variations of material microstructure may influence the backscattered grain noise. We developed new ideas of how to extract useful microstructural information from the forging simulation software *Simufact.forming*, a commercial software package produced by *Simufact Engineering GmbH*. A model is then developed to correlate the grain noise signals with the microstructural variations due to the inhomogeneous plastic deformation associated with the forging processing. The grain noise levels predicted by the model at various locations are compared with experiments. Reasonably good agreements are observed.

Keywords: ultrasonic grain noise, simulation software, elastic anisotropy.

1 INTRODUCTION

In principle, the noise caused by the structure of polycrystalline materials is the result of the inhomogeneity of the acoustic properties of the material. These in-homogeneities are primarily associated with the anisotropy of the elasticity constants in grains or crystals. Total scattered radiation can be controlled by grain morphology, grain orientation, and elastic anisotropy of the microstructure.

Large variations in material noise were observed in connection with the position and direction of examination for Ti–6Al–4V slabs. Such variations are supposed to be correlated with variations of the microstructure that come from the thermo-mechanical process.

The thermo-mechanical process is the most used way to control the microstructure and consequently the mechanical properties (for example imposing restrictions on the working temperature and the degree of deformation). A large variety of microstructural characteristics of the final product (grain size, degree of recrystallization, texture, etc.) are related to the parameters of thermo-mechanical processes.

Thus, in order to be able to anticipate the ultrasonic behaviour of a material, it is essential to anticipate its structural characteristics as a result of TMP, for which purpose we use computerized prediction techniques of the forging process.

WIT Transactions on Engineering Sciences, Vol 130, © 2021 WIT Press
www.witpress.com, ISSN 1743-3533 (on-line)
doi:10.2495/CMEM210121

2 MATERIAL NOISE PREDICTION ELEMENTS

A typical thermo-mechanical process is the one applied in the case of forging ingots (Fig. 1). The cast ingot is transformed into a cylindrical billet by reducing the diameter and then by forging it is brought to a shape as close as possible to the final product.

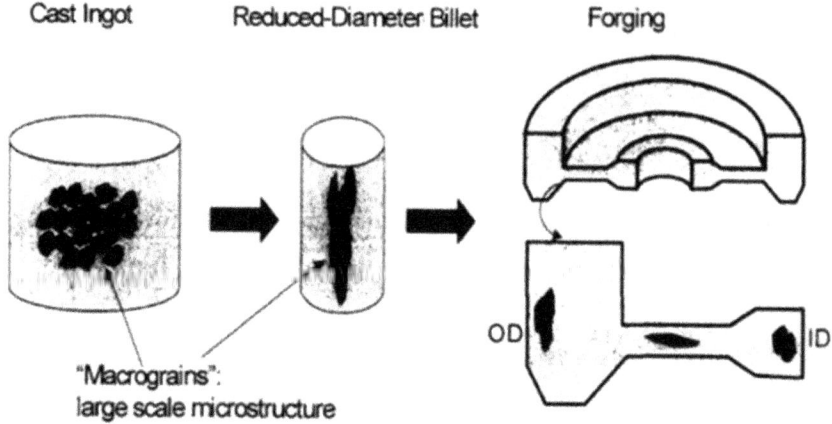

Figure 1:   Typical forging process accompanied by deformation and orientation of the grain depending on their position in the final product [1].

The microstructure of the ticket will depend on the microstructure of the casting and the history of the thermo-mechanical process. In the cast ingot, the grains are typically equiaxi, but due to the drastic reduction in diameter, the typical microstructure of the billet will be with elongated macro-grains (primary β), consisting of α colonies of hexagonal crystals created by solid state transformations with the orientation determined in the piece of that of the initial β grains. The final microstructure of the forging is dependent on the specifics of the forging process (forging temperature, processing time, deformation rate, cooling rate and subsequent heat treatment).

Apart from the morphology of the structure, the in-homogeneities of the plastic deformations during forging induce a preferential orientation of the crystals, a texture formed by the cleavage or maclation of the crystals, or a combination between the two.

Such deformations on which the texture depends will act as the elastic anisotropy factor. $<\delta C_{ij} \delta C_{kl}>$ vary with the position and direction inside the slab. These variations influence the material noise.

Due to these deformations, the deformation texture appears, which is conventionally represented by polar figures that describe the statistical distribution of the hexagonal basal plane. The extent and intensity of the texture are dependent on the volume of deformation and the temperature at which it was performed.

Due to the influence of inhomogeneous plastic deformation, the microstructure is expected to have local variations in grain morphology, grain orientation and texture.

However, for large slabs it is difficult to give photographic details of structural variations only through traditional metallographic approaches.

For this purpose there are specialized software packages such as Simufact.forming produced by *Simufact Engineering GmbH*. An example of using the program Simufact.forming is given in Fig. 2 where the case of a forged cylindrical ticket type is presented. The figure shows the "stress map" where the variations of the plastic deformation stresses are highlighted according to their position in the part, in a longitudinal section of the forged billet.

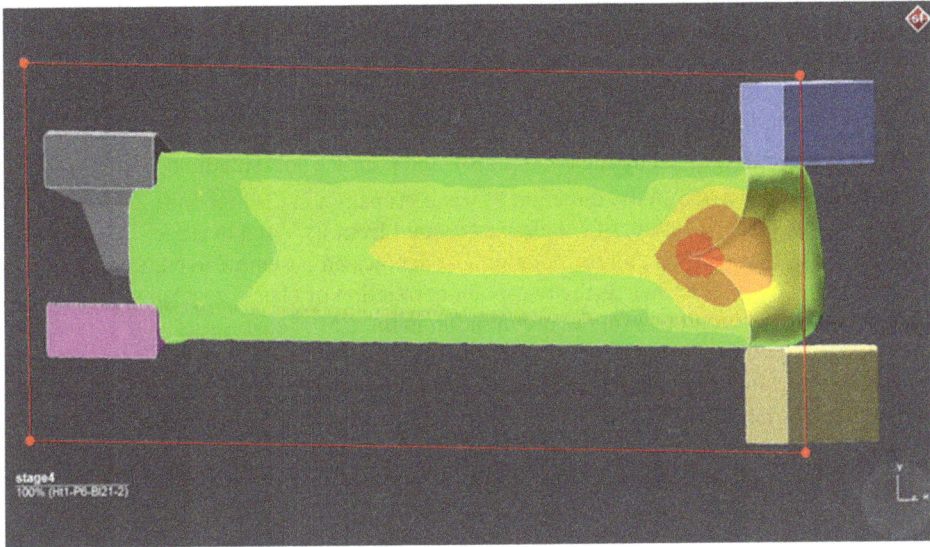

Figure 2:   Tension map in a forged ticket section.

An approach was thus developed through a mathematical model that can be predictable from the noise of material deduced from variations in microstructure, which in turn are dependent on plastic deformations during the forging process. It is possible to anticipate and thus determine the noise of the material and implicitly the quality of the material from the point of view of the ultrasonic control, in different areas of a forged part, by modelling the forging technology.

For this purpose, we start from the hypothesis of the elements that cause the scattering of ultrasonic radiation in the bill, having a specific orientation and elongation. These elements are a consequence of the forging process that changes their shape and orientation and introduces the deformation texture. The position and texture of these elements in the slab can be deduced by applying the program Simufact.forming.

2.1 The relationship between the elastic constants of two points located in a textured biphasic microstructure

To predict an absolute noise levels we need detailed knowledge of the metal microstructure which enters the model calculations through certain frequency-dependent factors known as "backscatter coefficients" or "Figures-of-Merit" (FOM). The resulting expression for "noise" produced under these conditions contains a factor which depends on the volumetric density of scatters and their Root Mean Square (RMS) scattering amplitude in the backscattered direction. This term, which is called FOM, appears to be a material property useful in characterizing microstructure. Under the assumption that the solid density is independent of position, FOM (ω) and η (ω) for longitudinal waves propagating in the S3 direction are related to microstructure features by eqn (1):

$$FOM\ (\omega)^{\ 2} = \eta\ (\omega) = K/(4\pi\rho V t^2)^2 \int d^3\vec{S}\ \langle\ \delta C_{33}(\ \vec{r}\ )\ \delta C_{33}(\vec{r'})\ \rangle e^{\,2iks3}, \tag{1}$$

where:

$\eta(\omega)$ – backscattering power coefficient,

ω – is the angular frequency,

k – is the magnitude of the wave vector of the incident wave,

ρ – is the density of the solid,

Vt – is the longitudinal wave velocity,

$S = \vec{r}\text{-}\vec{r'}$ – is a vector defined by the two points $\vec{r}$ and $\vec{r'}$ in the solid medium,

S_3 – is the component of $\vec{S}$ in the direction of wave propagation (3-direction),

$\delta C_{33}(\vec{r})$ – is the local deviation of the elastic constant from its Voigt average,

$\delta C_{33}(r^{\rightarrow}) - (C_{33} - C_{33}{}^{Voigt})$, and the notation $\langle...\rangle$ denotes an ensemble average, and $\langle \delta C_{33}(r^{\rightarrow})\ \delta C_{33}(r'^{\rightarrow})\rangle$ is known as the two-point correlation of elastic constants, which describes the correlation in the perturbation in the elastic stiffnesses.

For single-phase materials which are macroscopically homogeneous and have a random microstructure orientation, $\langle C33\ (r^{\rightarrow})\ \delta C33\ (r'^{\rightarrow})\rangle$ may be written in the form:

$$\langle \delta C_{33}\ (r^{\rightarrow})\ \delta C_{33}\ (r'^{\rightarrow})\ \rangle = <\delta C_{33}\ \delta C_{33}> W\ (r^{\rightarrow}\text{-}r'^{\rightarrow}), \tag{2}$$

where:

$<\delta C33\ \delta C33>$ – is a constant controlled by the crystal anisotropy;

$W\ (r^{\rightarrow}\text{-}r'^{\rightarrow})$ – the probability that two points separated by the distance $(r^{\rightarrow}\text{-}r'^{\rightarrow})$ in the solid will be in the same crystal;

$<\delta C33\ \delta C33>$ can be considered as an average of the solid angle (C33 – C33Voigt) for a single crystal.

$$FOM(\omega)^2 = \eta(\omega) = \left(\frac{\omega^2}{4\pi\rho v_l^2}\right)^2 < \delta C_{33}\ \delta C_{33} > \int d^3\vec{S}\ \ W(\vec{S})e^{2ikS_3}. \tag{3}$$

For the ellipsoidal element causing scattered radiation shown in Fig. 3, which is assumed to be axially symmetrical along the Z axis and the semi-axes a_m and c_m, eqn (3) can be written as in eqns (4)–(7).

$$FOM(\omega)^2 = \eta(\omega) = \left(\frac{\omega^2}{4\pi\rho v_l^2}\right)^2 < \delta C_{33}\ \delta C_{33} > \int_0^{2\pi}\int_0^{\pi}\frac{2\sin\theta}{A^3}\ d\theta d\Phi, \tag{4}$$

$$A = \frac{\{1+(R^2-1)\cos\theta^2\}^2}{(3VR/4\pi)^{1/3}} - 2ik\ (sin\ \theta\ sin\ \Phi\ sin\ \tau + cos\ \theta\ cos\ \tau), \tag{5}$$

$$R = \frac{a_m}{c_m}\ \text{– the ratio of the dimensions of the element causing the scattering and} \tag{6}$$

$$V = \frac{4\pi a_m^3}{3R}\ \text{– the average volume of the element causing the scattering.} \tag{7}$$

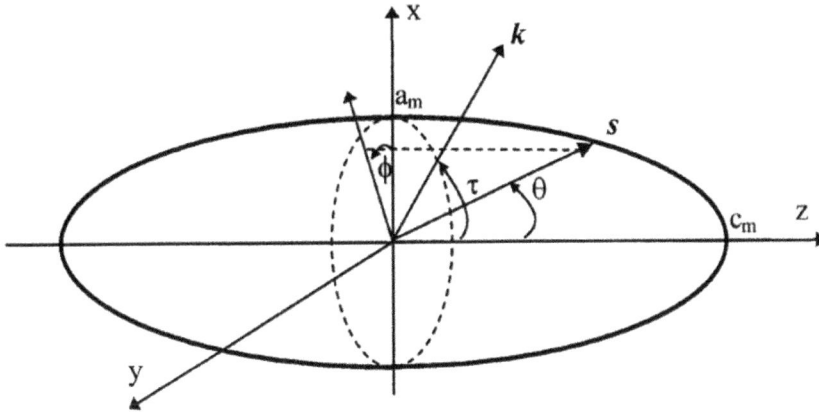

Figure 3:   Geometry of an ellipsoidal scattering element, sonic beam propagation is assumed to be parallel to vector k [1].

When the variations and intensity of the texture are small, its effect in correlating the elasticity functions between two points is given by the term $<\delta C33\delta C33>$. Due to the presence of texture, this term may change depending on the direction of examination.

The intensity and type of texture are defined as functions of orientation of the crystals, respectively by the probability that a certain crystal has a specific orientation in relation to the coordinate axes of the respective sample.

Specifically, the representation of a crystal in the microstructure is made by a set of Euler angles θ, ψ, ϕ. The orientation function of the crystals is represented by $w (\xi, \psi, \phi)$, where

$\xi = \cos \theta$ and $w (\cos \theta, \psi, \phi) \sin \theta \, d\theta \, d\psi \, d\phi$ is the probability of finding crystals oriented in the angular domain $d\theta \, d\psi \, d\phi$. [1]

Obviously,

$$\int_0^{2\pi} \int_0^{2\pi} \int_{-1}^1 w(\xi, \psi, \phi) \, d\xi \, d\psi \, d\phi = 1. \tag{8}$$

Once $w (\xi, \psi, \phi)$ is defined, the effect of the texture can be evaluated. In our case, $<\delta C33\delta C33>$ with texture can be evaluated with eqn (9) [2].

$$< \delta C_{33}\delta C_{33} > = \int_0^{2\pi} \int_0^{2\pi} \int_{-1}^1 [\delta C33(\xi, \psi, \phi)\delta C33(\xi, \psi, \phi)]w(\xi, \psi, \phi) \, d\xi \, d\psi \, d\phi. \tag{9}$$

The term $\delta C33 (\xi, \psi, \phi)$ in eqn (9) represents the local deviation from Voigt's average of the homogeneous alpha phase macrostructure, respectively $\delta C33 (\xi, \psi, \phi) = C33 (\xi, \psi, \phi) - C33$Voigt. Consider the basal plane of the crystals normal to the flow direction of the metal during plastic deformation, as shown in Fig. 4.

Mathematically, the function $w (\xi, \psi, \phi)$ is a function only of ξ or θ ($\xi = \cos \theta$). It is not possible to assess exactly to what extent the distribution of the basal poles changes with respect to θ. For reasons of mathematical correctness, $w (\xi)$ is expressed in the Gaussian form $w (\xi) = Ce^{-(\xi/\Delta)^2}$. By the normalization conditions established in eqn (8), the constant C is defined in eqn (10), where *erf* represents the function error [1].

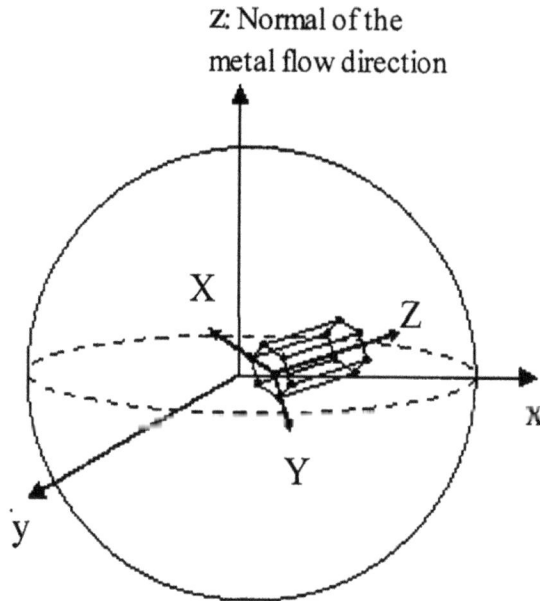

Figure 4: Crystallite alignment within forging [2].

$$w(\xi)) = \frac{1}{\left[4\pi^{5/2}\Delta erf\left(\frac{1}{\Delta}\right)\right]} e^{-(\xi/\Delta)^2}.$$ (10)

The term Δ is the one that determines the intensity of the texture. When Δ becomes very large, the texture disappears so that w (ξ) is a constant for a randomly distributed environment.

The small values of Δ correspond to a very pronounced texture. It is known that the intensity of the texture is determined by the volume of labor performed on the material or by the internal stresses resulting from its plastic deformations.

The relation between Δ and the internal tensions $\mathcal{E}e$ is given by eqn (11).

$$\Delta = \frac{P1}{\sqrt{\mathcal{E}e}},$$ (11)

where $P1$ is an adjustment parameter that can be deduced by comparing theoretical predictions with a number of experiments. Finally, the function w (ξ) can be expressed relative to the internal stresses $\mathcal{E}e$ according to eqn (12).

$$w(\xi) = \frac{1}{\left[4\pi^{5/2}P1 erf\left(\frac{\sqrt{\mathcal{E}e}}{P1}\right)\right]} e^{-\xi^2 \mathcal{E}e/P1^2}.$$ (12)

The effect of the texture $<\delta C33\delta C33>$ can be evaluated by replacing w (ξ) in eqn (9) with the expression from eqn (12). A specific example is given by Figs 5 and 6 in which a simulation performed using the program *Simufact.forming* 12.0., simulation program in the field of forging technologies is presented.

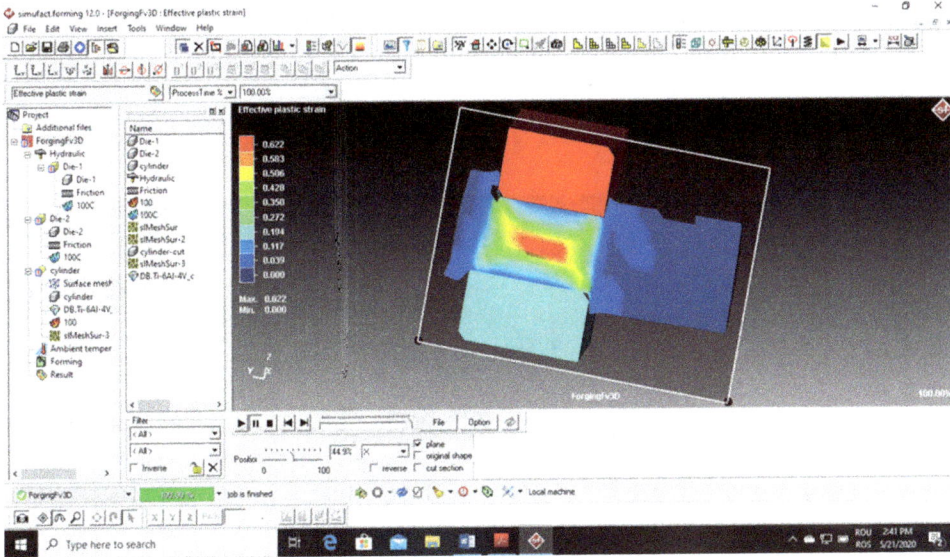

Figure 5: Plastic deformation stress map.

In this case, we started from a Ti6Al–4V ingot with a diameter of 400 mm and forged to a final diameter of 250 mm. Both the map of internal stresses and that of the material flow directions clearly indicate that these directions are rigorously aligned with the radial forging directions. As shown in Fig. 6, the basal plane of the crystals tends to align with the direction of flow of the metal. Since the values of the internal stresses can be evaluated with the help of the stress map obtained with the help of the *Simufact.forming* program, (Fig. 5) the exact value of $<\delta C33\delta C33>$ can be determined with eqn (12) and eqn (9) if *P1* is known. For biphasic structures ($\alpha + \beta$), the element $< \delta C_{ij} \delta C_{kl} >$ is smaller than that for pure α structures. To quantify this effect, a second adjustment parameter *P2* is entered.

$$< \delta C_{ij} \delta C_{kl} >_{\alpha+\beta} = P_2 < \delta C_{ij} \delta C_{kl} >_\alpha. \tag{13}$$

2.2 Calculation of the coefficient of power of the scattered radiation

We consider the model parameter V (eqn (7)) constant in the forging volume while the local parameter $<\delta C33\ \delta C33>_{\alpha + \beta}$ is determined by the two adjustment parameters *P1* and *P2* and by the plastic deformation stresses (eqns (9)–(13)). The calculation of the FOM is made for the direction of radial examination for tickets and normal to the major surface for forged plates. The values of the internal stresses $\mathcal{E}e$ (Fig. 5) and the orientation angles of the crystals in relation to the angle of incidence of the incident beam (Fig. 6) can be determined using the simulation program *Simufact.forming*, for a specific forging technology, at each point of the examined piece. The FOM calculation is done in several stages, as follows:

Stage 1 – A *FOM-Frequency Dependence Curve* is experimentally raised using the stepped reference block provided with Flat Bottom Holes (FBH) of Φ 1.2 mm and Φ 2.0 mm, positioned at different depths (Fig. 7). To do this, successively scan the FBH Φ 1.2 mm at a depth of 129.5 mm from the inlet surface, with three immersion transducers having the same dimension of the piezoelectric element but different central frequencies (respectively 2.25 MHz, 5 MHz and 10 MHz) and the oscillograms of each scan are recorded (Figs 8–10).

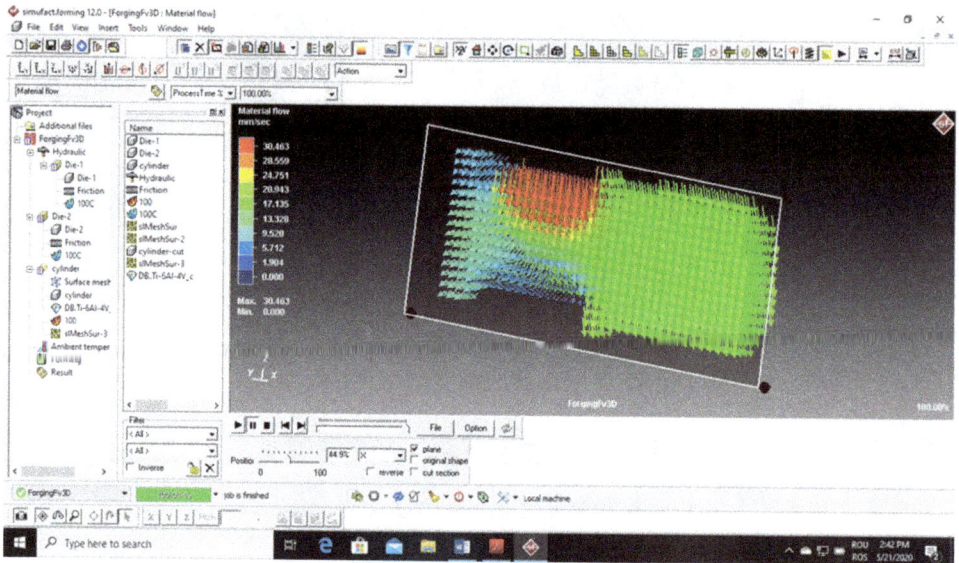

Figure 6: Map of speeds and flow directions of the material.

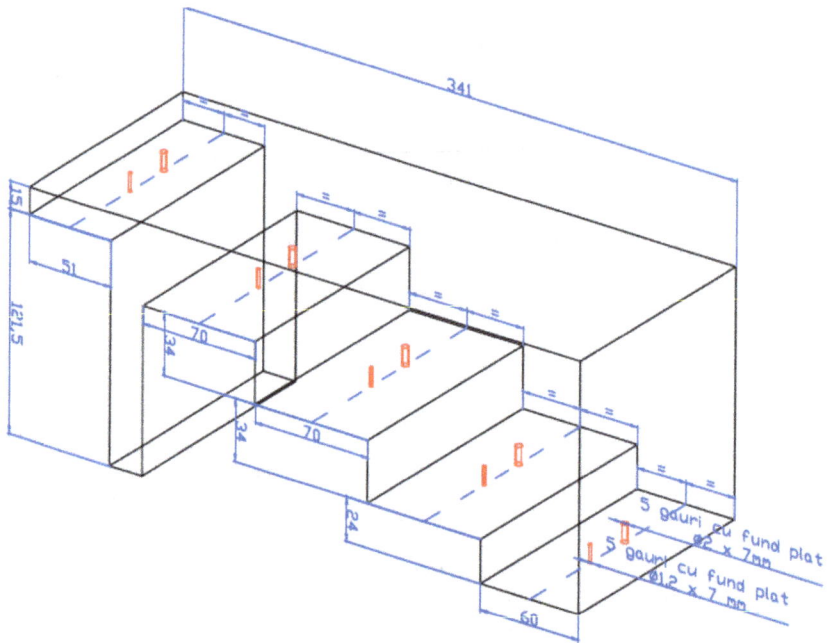

Figure 7: Forged plate examination reference block.

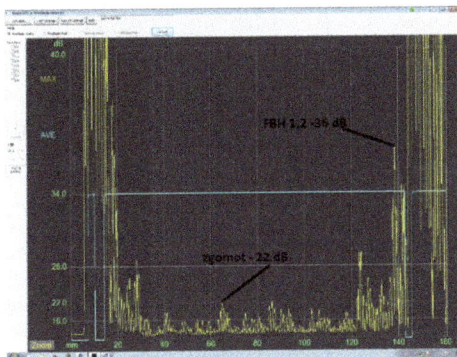

Figure 8: Transducer Φ 19 mm, 2.25 MHz.

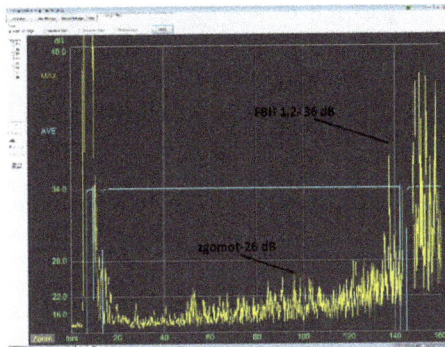

Figure 9: Transducer Φ19 mm, 5 MHz.

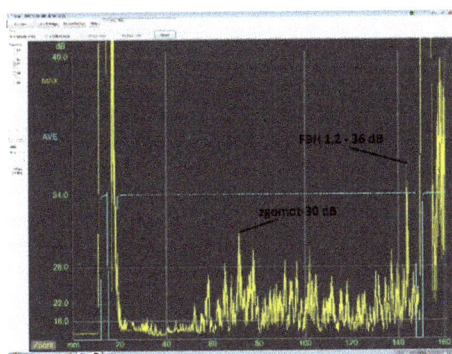

Figure 10: Transducer Φ19 mm, 10 MHz.

The examination sensitivity will be set so that the reference reflector is at the same relative amplitude, respectively FBH Φ 1.2 mm echo at 60% of Full Screen Height (FSH) for each scan. The material noise level (which is directly proportional to the FOM) will be recorded for each transducer and the *S/N* ratio for each transducer will be determined. Experimentally, these value of the *S/N* ratio are determined as follows:

- 2.25 MHz Transducer – S/N = 6:1
- 5.0 MHz Transducer – S/N = 4:1
- 10.0 MHz Transducer – S/N = 2:1

The *Signal/Noise* ratio can be expressed as a function of the pulse volume and the power coefficient of the scattered radiation η (ω) according to eqn (14) [2].

$$S/N = \left[\frac{A flaw(\omega)}{\sqrt{\eta(\omega)}} \right] \frac{1}{\sqrt{B^2 \, \Delta tp}}, \tag{14}$$

where:
- B – average ultrasonic pulse diameter,
- Δtp – ultrasonic pulse length and
- *A flaw* – amplitude of the reference defect at 60% FSH.

To determine the pulse volume, a FBH Φ1.2 mm located at a depth of 129.5 mm is used for the pulse diameter and the bottom echo from the lower surface of the step of 136 mm (Fig. 7) to determine the pulse length [3].

Given these data and the amplitude of the reference defect $Aflaw$, the same (60% FSH) for all examinations, we can determine the conversion factor between FOM and S/N ratio for each transducer, respectively each examination frequency. The dependency graph between FOM and frequency can now be drawn (Fig. 11).

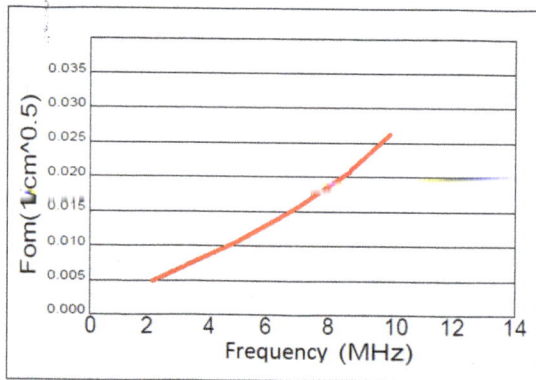

Figure 11: Dependency *FOM-Frequency*.

Stage 2 – The *Simufact.forming* program will be run and the degree of deformation of the material, volume V and ratio R will be set as initial values in order to obtain the values of the plastic deformation stresses εe for the scanned reference block.

Stage 3 – An FOM value will be appreciated from the graph in Fig. 11 for the frequency with which the examination is performed. Using the eqns (3)–(7), from this value determined graphically we can deduce the value $<\delta C33\delta C33>$ from which we can later deduce the value Δ, from eqn (11) and knowing the value εe deduced in *Stage 2*, the value of the adjustment parameter $P1$ can be deduced.

Stage 4 – Having the value of the adjustment parameter $P1$, and with the help of the simulation program *Simufact.forming* from which the angle of inclination of the basal plane is extracted in relation to the incident beam τ, volume V and ratio R, we can calculate FOM (eqn (4)) for any forged part like the one in Fig. 12, at any point thereof. Since the FOM value is also related to the S/N ratio, for the same transducer and on approximately equal working fields, it can be evaluated by comparing the FOM values obtained by calculation on scan areas 1, 2 and 3 (Fig. 12) with the value obtained on the reference block, which from the areas of the plate to be examined meets the S/N ratio conditions required by the standard.

Another element of material noise prediction is the orientation of the crystals according to the scanning direction. This orientation is related to the direction of flow of the material (Fig. 13) and can be anticipated in order to determine the defining angles for the element causing the scattering (Fig. 3), using the same simulation program. The forging technology will be modified accordingly in order to achieve those material and texture conditions that allow the efficient ultrasonic examination of the semi-finished product.

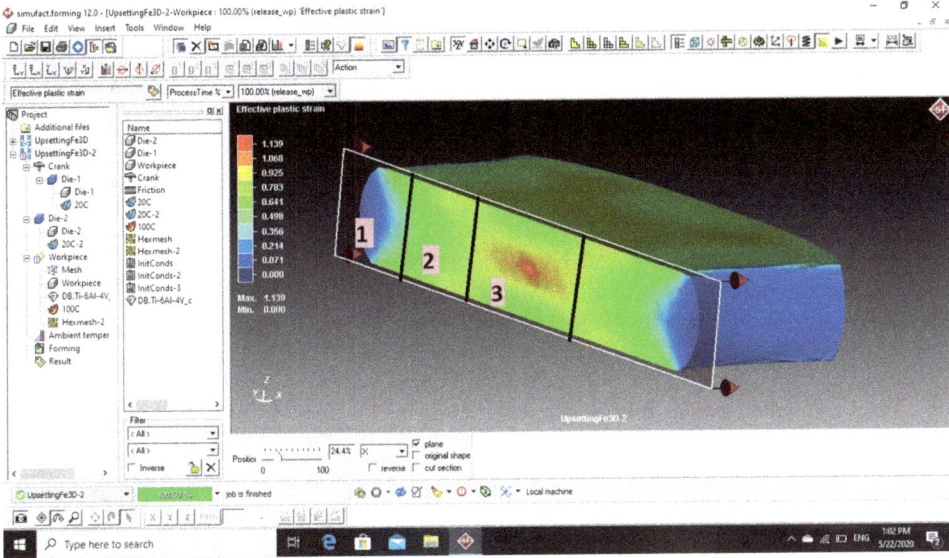

Figure 12: Map of plastic deformation stresses in a forged plate.

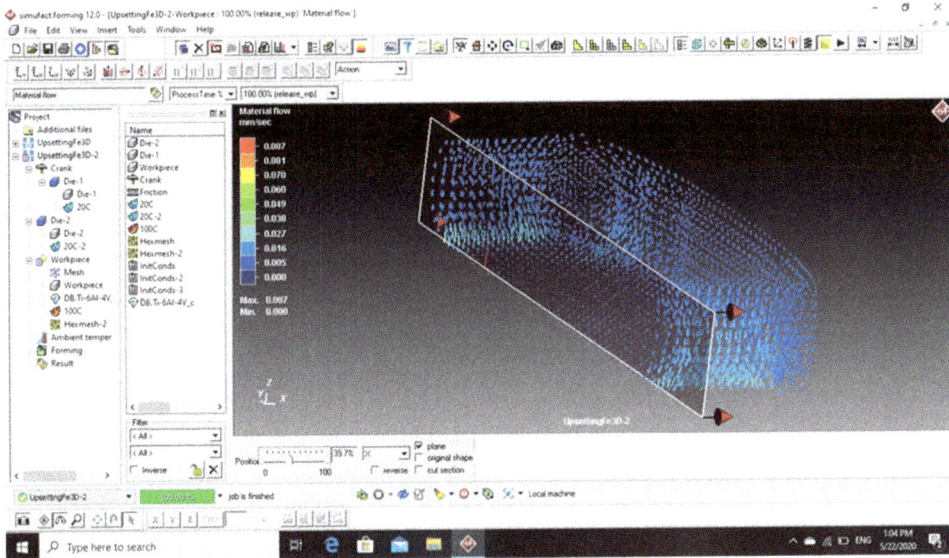

Figure 13: Map of velocities and flow directions of the material in a forged plate.

3 CONCLUSIONS

We set out to develop a model that correlates the material noise with the microstructural variations resulting from the plastic deformations during forging. The inputs of this model refer to the average volume V of the element that causes the scattering, supposed to be constant during forging, to the geometric characteristics of this element and to the local texture due to the inhomogeneity of the plastic deformation stresses.

The shape and orientation of the elements causing the scattering can be deduced with the *Simufact.forming* simulation program and vary over the entire volume of the slab. The other parameters necessary for modeling, respectively the adjustment parameters *P1* and *P2* for the characterization of the anisotropy of the elastic properties were obtained experimentally.

The aim of this paper is a better understanding of the fundamental ultrasonic properties of Ti–6A1–4V slabs as well as the improvement of inspection and manufacturing procedures of this material.

ACKNOWLEDGEMENTS
This work was supported by S.C. ZIROM-S.A – Giurgiu – Romania and was performed by the NDT Consulting Company – DIAC SERVICII srl and AROEND – Romania.

REFERENCES
[1] Yu, L., *Understanding and Improving Ultrasonic Inspection of Jet-Engine Titanium Alloy,* Iowa State University, 2004.
[2] Van Pamel, A., *Ultrasonic Inspection of Highly Scattering Materials,* Imperial College London, Department of Mechanical Engineering, 2015.
[3] Margetan, F.J., Umbach, J., Roberts, R., Friedl, J. & Degtyar, R., *Inspection Development for Titanium Forgings,* DOT/FAA/AR-05/46, May 2007.

Author index

WIT*PRESS* ...for scientists by scientists

Multiphase Flow: Computational & Experimental Methods

Edited by: **S. HERNÁNDEZ**, *University of A Coruña, Spain and* **P. VOROBIEFF**, *University of New Mexico, USA*

The research included in this volume focuses on using synergies between experimental and computational techniques to gain a better understanding of all classes of multiphase and complex flow. The included papers illustrate the close interaction between numerical modellers and researchers working to gradually resolve the many outstanding issues in our understanding of multiphase flow.

Recently multiphase fluid dynamics have generated a great deal of attention, leading to many notable advances in experimental, analytical and numerical studies. Progress in numerical methods has permitted the solution of many practical problems, helping to improve our understanding of the physics involved.

Multiphase flows are found in all areas of technology and the range of related problems of interest is vast, including astrophysics, biology, geophysics, atmospheric process, and many areas of engineering.

ISBN: 978-1-78466-417-6   **eISBN: 978-1-78466-418-3**

Published 2020 / 136pp